The Mathematics of GAMES OF STRATEGY

Theory and Applications

by

Melvin Dresher

Research Mathematician
The Rand Corporation

Dover Publications, Inc., New York

To my brothers
Isaac, Jacob, and Charles

Published in Canada by General Publishing Company, Ltd., 30 Lesmill Road, Don Mills, Toronto, Ontario.
Published in the United Kingdom by Constable and Company, Ltd.

This Dover edition, first published in 1981, is an unabridged republication of the work originally published in 1961 by Prentice-Hall, Inc., Englewood Cliffs, N.J., under the title *Games of Strategy: Theory and Applications*. A new Preface written by the author, superseding the original one, was provided especially for this edition, which is published by special arrangement with The Rand Corporation.

International Standard Book Number: 0-486-64216-X
Library of Congress Catalog Card Number: 81-68485

Manufactured in the United States of America
Dover Publications, Inc.
180 Varick Street
New York, N.Y. 10014

PREFACE

This book presents the mathematical theory of games of strategy and some of its applications. The mathematical theory is presented in an elementary and formal manner and the applications are of two kinds: games in the ordinary sense and competitive situations in general that exist in economics, military, and behavioral sciences. It is a book about decision making in the absence of perfect information. In particular, we are concerned with decision problems in a competitive environment where conflicting interests exist, and uncertainties and risk are involved.

The mathematical presentation is elementary in the sense that no advanced algebra or non-elementary calculus occurs in most of the mathematical proofs. In the few instances requiring advanced algebra or more than basic calculus, the advanced topic is developed prior to its use in the proof.

The applications presented deal with some decision problems in economics, military, business, and operations research. Although most of the applications are discussed in military terminology using such terms as attack and defense, they can be formulated in economic or behavioral-science terms. Further, many games of strategy, other than military, are frequently described in military language. Such military terms as "bluffing," "opponent's capabilities," "opposing courses of action," often appear in the vocabulary of non-military fields of competition.

We have attempted to present game theory as a branch of applied mathematics. In addition to developing a mathematical theory for solving games, we show how to formulate a game model associated with a given competitive or conflicting situation. We also show how some decision problems, such as timing of decisions, which do not resemble game situations can be analyzed as a game, yielding rich insight into the decision problem.

Each application is introduced by first describing the competitive or conflicting situation that is to be analyzed. Next, we describe the simplifications that are to be made, such as reducing the number of choices. Then we give the abstract or mathematical formulation of the strategic situation, yielding the game and the associated payoff, which we proceed to analyze mathematically. The mathematical analysis frequently yields results which are unexpected and provides an understanding of the competitive environment.

An attempt has been made to develop the subject matter in such a way as to make the volume adaptable as a text on the theory of games in colleges and universities. The analytical level of the book is such that one year of calculus can serve as a prerequisite for the course.

The book starts in Chapter 1 with an exposition of games of strategy, with examples taken from parlor games as well as from military games. The next two chapters treat the basic topics in the theory of finite games, i.e., the existence of optimal strategies and their properties. We give an elementary proof of the minimax theorem which also provides an efficient method for computing optimal strategies. Chapters 4 and 5 deal with the representation of games in extensive form and methods of computing optimal stratetgies.

Since many games involve an infinite number of strategies, Chapters 6, 7, and 8 deal with such games by first developing the necessary mathematics (e.g., probability distribution functions and Stieltjes integrals) for analyzing infinite games. The existence of optimal strategies and their properties are discussed in detail. We give an example of an infinite game which does not have a solution.

The results on infinite games are applied in Chapters 9 and 10 to two general classes of games—timing games and tactical games. Finally, the last chapter provides an application of moment space theory to the solution of infinite games.

Many of the chapters are independent of one another; for example, the chapter on games in extensive form, Chapter 4, may be introduced after Chapter 1, while the last four chapters may be studied in any order after Chapter 7.

This study was undertaken by The Rand Corporation as a part of its research program for the United States Air Force.

An author of a textbook is indebted to everyone who has made a contribution to the subject. In particular, I wish to express my indebtedness to the authors of books and papers cited in the Bibliography. All of them have contributed to my appreciation and understanding of the theory of games.

Many of the applications and examples discussed in this book are due to people other than the author. I am grateful for specific applications to R. E. Bellman, R. L. Belzer, L. D. Berkovitz, O. Gross, O. Helmer, S. M. Johnson, E. W. Paxson, and L. S. Shapley, all of The Rand Corporation; and also to D. Blackwell of the University of California, H. W. Kuhn, Oskar Morgenstern, and A. W. Tucker of Princeton University, S. Karlin of Stanford University, and Rufus Isaacs of Johns Hopkins University.

I wish to thank Samuel M. Genensky, who read the manuscript. Thanks are due to Dorothy G. Stewart for patient and expert help in proofreading.

June 1980 *Melvin Dresher*
Santa Monica, California

TABLE OF CONTENTS

1 GAME, STRATEGY, AND SADDLE-POINT

1. INTRODUCTION

The theory of games of strategy may be described as a mathematical theory of decision-making by participants in a competitive environment. In a typical problem to which the theory is applicable, each participant can bring some influence to bear upon the outcome of a certain event; no single participant by himself nor chance alone can determine the outcome completely. The theory is then concerned with the problem of choosing an optimal course of action which takes into account the possible actions of the participants and the chance events.

Examples of games of strategy are parlor games such as poker, chess, and bridge, military games such as the defense of targets against attack, and economic games such as the price competition between two sellers. Each of these games of strategy allows the players to make use of their ingenuity in order to influence the outcome. Although some of these games involve elements of chance (e.g., the cards dealt in poker and the hit of a target), we shall exclude from our discussion games of chance such as blackjack and craps since their outcomes depend entirely upon chance and cannot be influenced by the decisions of the players.

Games of chance have been studied mathematically for many years; the mathematical theory of probability was developed from their study. Although strategic situations have long been observed and recorded, the first attempt to abstract them into a mathematical theory of strategy was made in 1921 by Émile Borel. The theory was firmly established by John von Neumann in 1928 when he proved the minimax theorem, the fundamental theorem of games of strategy. However, it was not until the publication in 1944 of the impressive work, *Theory of Games and Economic*

Behavior, by John von Neumann and Oskar Morgenstern, that the mathematical theory of games received much attention. This book emphasized a new approach to the general problem of competitive behavior through a study of games of strategy. Since then the theory has been applied to game-like problems—not only in economics, but also in the military and politics.

2. DESCRIPTION OF A GAME OF STRATEGY

A game of strategy is described by its set of rules. These rules specify clearly what each person called a *player* is allowed or required to do under all possible circumstances. The rules define the amount of information, if any, each player receives. If the game requires the use of chance devices, the rules describe how the chance events shall be interpreted. They also define the time the game ends, the amount each player pays or receives, and the objective of each player.

From the rules we can obtain such general properties of the game as the number of moves, the number of players, and the payoff. The game is *finite* if each player has a finite number of moves and a finite number of choices available at each move. It is convenient to classify games according to the number of players, i.e., as 2-person, 3-person, etc. It is also convenient to distinguish between games whose payoffs are zero-sum and those whose are not. If the players make payments only to each other, the game is said to be *zero-sum*.

We shall first analyze the 2-person, zero-sum, finite game—most parlor games and many military games are of this type. It is clear that in such a game the winnings of one player are the losses of the other player.

To simplify the mathematical description of a game, we introduce the concept of a "strategy." In the actual play of a game, instead of making his decision at each move each player may formulate in advance of the play a plan for playing the game from beginning to end. Such a plan must be complete and cover all possible contingencies that may arise in the play. Among other things, this plan would incorporate any information which may become available to the player in accordance with the rules of the game. Such a complete prescription for a play of a game by the player is called a *strategy* of that player. A player using a strategy loses no freedom of action since the strategy specifies the player's actions in terms of the information that might become available.

We may think of a strategy of a player as a set of instructions for playing the given game from the first move to the last. Conversely, each different way that a player may play a game is a strategy of that player. If we enumerate all the different ways for a player to play the game, we obtain all the strategies of that player. Of course, many of these strategies may be obviously poor ones, but they are included in the initial enumeration.

Every pair of strategies, consisting of one strategy for each player, determines a play of the game, which in turn determines a payoff to each player. Thus we may consider a play of a game to consist of each player's making one decision, namely the selection of a strategy.

In terms of this notion of strategy, we can now describe, corresponding to any given game, an equivalent game of much simpler character. Let us call the two players Blue and Red. Suppose Blue has m strategies, which may be designated by the numbers

$$i = 1, 2, \ldots, m.$$

Suppose Red has n strategies, which we may designate by

$$j = 1, 2, \ldots, n.$$

Then on the first move Blue chooses some strategy i. On the next move, Red, without being informed what choice Blue has made, chooses strategy j. These two choices determine a play of the game and a payoff to the two players. Let a_{ij} be the payoff to Blue. Since the game is zero-sum, the payoff to Red is $-a_{ij}$.

The game is thus determined by Blue's payoff matrix,

$$A = \begin{bmatrix} a_{11} & a_{12} & \cdots & a_{1n} \\ a_{21} & a_{22} & \cdots & a_{2n} \\ \cdot & \cdot & \cdots & \cdot \\ \cdot & \cdot & \cdots & \cdot \\ \cdot & \cdot & \cdots & \cdot \\ a_{m1} & a_{m2} & \cdots & a_{mn} \end{bmatrix}.$$

In this matrix each Blue strategy is represented by a row; each Red strategy is represented by a column. If Blue chooses the ith strategy or ith row and Red chooses the jth strategy or jth column, then Red is to pay Blue the amount a_{ij}. Blue wants a_{ij} to be as large as possible, but he controls only the choice of his strategy i. Red wants a_{ij} to be as small as possible but he controls only the choice of his strategy j. What are the guiding principles which should determine the choices and what is the expected outcome of the game?

3. ILLUSTRATIVE EXAMPLES

EXAMPLE 1. FUNCTION OF A FIELD COMMANDER.

Military action by a commander against an enemy requires him to evaluate a situation and make a decision. The evaluation is made by considering the given mission, the capabilities of the enemy, and the possible courses of action available to both sides, and by comparing the likely consequences of one's own courses of action. These are the rules of the game.

From his possible courses of action, the commander selects an optimal one—a course which promises to be most successful in accomplishing the mission.

The military doctrine covering the above evaluation is discussed in detail in the *Naval Manual of Operational Planning* of the Naval War College. The *Manual* states, "Each of our own courses of action . . . is separately weighed in turn against each capability of the enemy which may interfere with the accomplishment of the mission. The results to be expected in each case are visualized. The advantages and disadvantages noted as a result of the analysis for each of our own courses of action are summarized and the various courses of action are compared and weighed." A commander thus enumerates the possible opposing courses of action and their outcomes. Each course of action is a strategy for the commander and each outcome yields a payoff.

Suppose that the mission of the Blue forces is to capture an objective defended by the Red forces. The Blue commander analyzes the possible courses of action within the capabilities of the Red forces which can affect the capture of the objective. Then Blue lists the various ways of capturing the objective. Let us assume that this yields three possible courses of action, r_1, r_2, r_3, to the Red commander and three possible courses of action, b_1, b_2, b_3, to the Blue commander. There are then nine possible outcomes which may be summarized by a 3×3 matrix. For example, the matrix may have the following descriptive entries:

"ESTIMATE OF THE SITUATION" BY BLUE COMMANDER

$$\begin{array}{ll} & \qquad\qquad \textit{Red courses of action} \\ & \qquad \begin{array}{ccc} r_1 & \quad r_2 & \quad r_3 \end{array} \\ \textit{Blue courses of action} & \begin{array}{c} b_1 \\ b_2 \\ b_3 \end{array} \begin{bmatrix} \text{Fail} & \text{Succeed} & \text{Succeed} \\ \text{Draw} & \text{Succeed} & \text{Draw} \\ \text{Succeed} & \text{Draw} & \text{Fail} \end{bmatrix} \end{array}$$

In this illustrative situation, the Blue commander estimates that if he uses strategy b_1, for example, and the Red commander uses r_1, Blue will fail to capture the objective. However, Blue will succeed if he uses strategy b_1 and the Red commander uses strategy r_2 or r_3. Similarly, the remaining six outcomes are the Blue commander's evaluation of the outcomes with respect to various courses of actions.

The problem of the Blue commander is to select the best course of action, or the strategy which will secure for him the most likely chance of capturing the objective. The Red commander will likewise try to select his best course of action, or that strategy which will minimize Blue's chance of capturing the objective. We thus have the following problem in competitive behavior: Blue wishes to maximize the outcome by a proper choice of his strategy, and Red wishes to minimize this same outcome, also by a proper choice of Red's strategy, where the outcome depends upon both Blue's and Red's choices.

The established doctrine of the "Estimate of the Situation" dictates the selection of the course of action which promises to be most successful in the accomplishment of the mission regardless of the course selected by the Red commander. Accordingly, Blue would choose, in this illustrative example, strategy b_2 which yields either a successful outcome or a draw. For if the Blue commander uses a strategy b_1 or b_3 he may fail to accomplish the mission. Strategy b_2 represents the most conservative decision of the Blue commander. It is equivalent to assuming that the Red commander can find out the decision of the Blue commander. In a later section we shall see how the Blue commander can, by taking chances, or bluffing, obtain on the average a more advantageous outcome. He accomplishes this by mixing his strategies, or by choosing his strategy with a chance device.

EXAMPLE 2. THE GAME "MORRA."

This is a game which has only personal moves. Each player shows one, two, or three fingers and simultaneously calls his guess of the number of fingers his opponent will show. If just one player guesses correctly, he wins an amount equal to the sum of the fingers shown by himself and his opponent; otherwise the game is a draw.

This game consists of one move for each player—the choice of a number to show and a number to guess. We may represent a strategy for each player by a pair of numbers (s, g), where $s = 1, 2, 3$ is the number of fingers he shows and $g = 1, 2, 3$ is his guess of the number of fingers his opponent will show. It is evident that each player has nine strategies, and thus there are 81 different possible plays of the game. With each of these 81 ways of playing the game, there is associated a payment to the players, as described by the rules of the game. These payments are summarized by a payoff matrix. In the following payoff matrix, the entries represent payments to Blue. Red will receive the negative of these payments.

MORRA PAYOFF

Blue strategies \ *Red strategies*	(1, 1)	(1, 2)	(1, 3)	(2, 1)	(2, 2)	(2, 3)	(3, 1)	(3, 2)	(3, 3)
(1, 1)	0	2	2	−3	0	0	−4	0	0
(1, 2)	−2	0	0	0	3	3	−4	0	0
(1, 3)	−2	0	0	−3	0	0	0	4	4
(2, 1)	3	0	3	0	−4	0	0	−5	0
(2, 2)	0	−3	0	4	0	4	0	−5	0
(2, 3)	0	−3	0	0	−4	0	5	0	5
(3, 1)	4	4	0	0	0	−5	0	0	−6
(3, 2)	0	0	−4	5	5	0	0	0	−6
(3, 3)	0	0	−4	0	0	−5	6	6	0

EXAMPLE 3. THE GAME "LE HER."

This game involves both chance and personal moves. Although the game can be played by several people, we shall describe it for two players. From an ordinary deck of cards, a dealer gives a card at random to a receiver and takes one himself; neither player sees the other's card. The main object is for each to obtain a higher card than his opponent. The order of value is ace, two, three, . . . , ten, jack, queen, king. Now if the receiver is not content with his card, he may compel the dealer to change with him; but if the dealer has a king, he is allowed to retain it. If the dealer is not content with the card which he at first obtained, or which he has been compelled to take from the receiver, he is allowed to change it for another taken out of the deck at random; but if the card he then draws is a king, he is not allowed to have it, but must keep the card which he held after the receiver exercised his option. The two players then compare cards, and the player with the higher card wins. If the dealer and receiver have cards of the same value, the dealer wins.

This game consists of three moves: The first move is a chance move, one card each being dealt at random to the receiver and dealer. The second move is a personal move by the receiver, who exchanges his card with the the dealer or stays with the original card. The third move is a personal move by the dealer, who exchanges his card with a card from the deck or stays with the card he holds.

The game can be summarized by defining a strategy for the receiver to be a determination of change or stay for each of the thirteen cards he may receive. One such strategy might be

$$\begin{matrix} 1 & 2 & 3 & 4 & 5 & 6 & 7 & 8 & 9 & 10 & J & Q & K \\ [C & S & C & S & S & C & C & S & S & C & S & C & S] \end{matrix}$$

where "C" means change and "S" means stay. Thus, the above strategy tells the receiver to change if he receives an ace, stay if he receives a two, change if he receives a three, . . . , change if he receives a queen, stay if he receives a king. Note that the strategy is a complete set of instructions. It is apparent that the receiver has 2^{13} strategies. Of course, most of them are poor strategies and should not be played, as will be described in a subsequent section.

Since the dealer's personal move, the third move of the game, is made after the receiver has exercised his option, the dealer may then have information about the receiver's card. Therefore a strategy for the dealer must include this information. However, to simplify the enumeration of the dealer's strategy, let us make the following two observations about rational behavior of the dealer:

(i) If the receiver exchanges cards with the dealer and the dealer thereby obtains a lower card than the receiver, then the dealer should exchange with the deck.

(ii) If the receiver exchanges cards with the dealer and the dealer thereby obtains a card at least as high as the receiver's, then the dealer should stay with that card.

Let us refer to these two instructions of rational behavior as R. They have the effect of eliminating obviously poor strategies. Now the strategies of the dealer may be enumerated in a manner similar to those of the receiver and at the same time incorporate the information about the receiver. A strategy for the dealer may be written as follows:

$$(\mathrm{R};\ \mathrm{S\,C\,C\,S}\ldots\mathrm{S\,C}).$$

This strategy tells the dealer to do the following: if the receiver exchanges cards, the dealer follows the two instructions R of rational behavior; if the receiver stays with his card and the dealer holds an ace, then the dealer stays; if the receiver stays and the dealer holds a two, then the dealer changes with the deck; etc. The dealer now has 2^{13} strategies. Again, most of these strategies will turn out to be poor strategies as will be described in a later section.

EXAMPLE 4. COLONEL BLOTTO GAME.

Colonel Blotto and his enemy each try to occupy two posts by properly distributing their forces. Let us assume that Colonel Blotto has 4 regiments and the enemy has 3 regiments which are to be divided between the two posts. Define the payoff to Colonel Blotto at each post as follows: If Colonel Blotto has more regiments than the enemy at the post, Colonel Blotto receives the enemy's regiments plus one (the occupation of the post is equivalent to capturing one regiment); if the enemy has more regiments than Colonel Blotto at the post, then Colonel Blotto loses one plus his regiments at the post; if each side places the same number of regiments, it is a draw and each side gets zero. The total payoff is the sum of the payoffs at the two posts.

Colonel Blotto has 5 strategies, or five different ways of dividing 4 regiments between the two posts. The enemy has 4 strategies, or four different ways of dividing his 3 regiments. There are therefore twenty ways for the two sides to distribute their forces.

It is evident that if Colonel Blotto places 3 regiments at the first post and 1 at the second, and if the enemy places 2 regiments at the first post and 1 at the second, then Blotto wins what amounts to 3 regiments.

However, if Colonel Blotto places 2 regiments at each post and the enemy places all of his 3 regiments at either post, then Colonel Blotto loses 2 regiments. The following payoff matrix summarizes the payment to Colonel Blotto for each of the twenty possible distributions:

COLONEL BLOTTO PAYOFF

		Enemy strategies			
		(3, 0)	(0, 3)	(2, 1)	(1, 2)
Colonel Blotto strategies	(4, 0)	4	0	2	1
	(0, 4)	0	4	1	2
	(3, 1)	1	−1	3	0
	(1, 3)	−1	1	0	3
	(2, 2)	−2	−2	2	2

EXAMPLE 5. A COIN GUESSING GAME.

Two players secretly choose to conceal 0, 1, or 2 coins. Each, in agreed turn, tries to guess exactly the total number concealed by the two players. However, the second guesser must guess a different number from that called by the first. If a player guesses correctly, he receives 1 from the other; otherwise 0.

A strategy involves a joint decision of what to conceal and what to call. We may represent a strategy of Blue by a vector $(x, x + y)$ where $x = 0$, 1, or 2 and $y = 0$, 1, or 2. The first component, x, of the vector represents the number to conceal; the second component, $x + y$, represents the number to call. Thus Blue has 9 strategies.

A strategy for the player guessing second, call him Red, must include the information about Blue's guess, which will be 0, 1, 2, 3, or 4. We can represent a Red strategy by a vector,

$$(H;\ x_0, x_1, x_2, x_3, x_4),$$

where $H = 0, 1, 2$ is the number of coins concealed and x_i is the number of coins to call if Blue calls i. Red has 10 strategies. They are enumerated as follows:

$$\begin{array}{lll} (1) = (0;\ 10012), & (2) = (0;\ 10022), & (3) = (0;\ 10112), \\ (4) = (0;\ 10122), & (5) = (1;\ 12123), & (6) = (1;\ 12323), \\ (7) = (2;\ 22343), & (8) = (2;\ 22443), & (9) = (2;\ 23343), \\ & (10) = (2;\ 23443). & \end{array}$$

The payoff matrix is now readily written. It is

Coin Game Payoff

		Red strategies									
		(1)	(2)	(3)	(4)	(5)	(6)	(7)	(8)	(9)	(10)
Blue strategies	(0, 0)	1	1	1	1	−1	−1	−1	−1	−1	−1
	(0, 1)	−1	−1	−1	−1	1	1	−1	−1	0	0
	(0, 2)	−1	−1	0	0	−1	0	1	1	1	1
	(1, 1)	1	1	1	1	−1	−1	0	0	−1	−1
	(1, 2)	0	0	−1	−1	1	1	−1	0	−1	0
	(1, 3)	−1	0	−1	0	−1	−1	1	1	1	1
	(2, 2)	1	1	1	1	0	−1	0	−1	0	−1
	(2, 3)	0	−1	0	−1	1	1	−1	−1	−1	−1
	(2, 4)	−1	−1	−1	−1	−1	−1	1	1	1	1

Example 6. Simplified Poker.

Each of two players antes one unit. They obtain a fixed hand by drawing one card apiece from a pack of three cards numbered J, Q, K. Then the players choose alternately to bet one unit or to pass without betting. Two successive bets or passes terminate a play, at which time the player holding the higher card wins the amount previously wagered by the other player. A player passing after a bet also ends a play and loses his ante.

There are six possible deals. For each deal the plays of the game may be diagrammed as shown in Fig. 1. We see that for each deal there are five plays. Thus, thirty possible plays exist in this game.

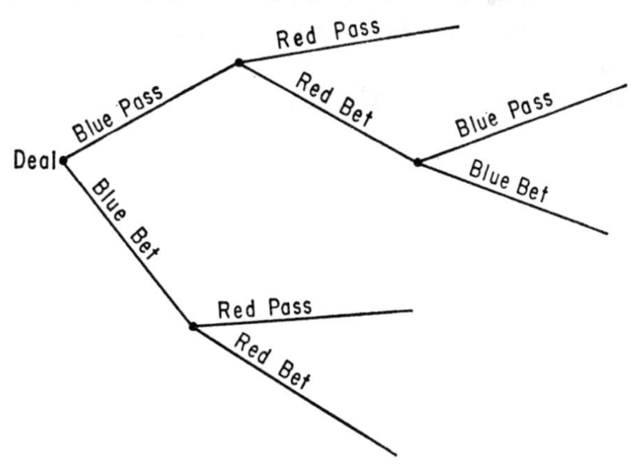

Figure 1

A strategy for the player choosing first, call him Blue, must tell him whether to pass or bet for each of the three possible cards he may hold.

A convenient way to code Blue's strategy is by a vector (a_J, a_Q, a_K)

where a_i gives the instructions for card i = J, Q, K. Then a_i may be coded as follows:

$$a_i = \begin{cases} 0 & \text{means pass in round 1 and pass in round 2} \\ 1 & \text{means pass in round 1 and bet in round 2} \\ 2 & \text{means bet in round 1.} \end{cases}$$

It is clear that Blue has $3 \times 3 \times 3 = 27$ strategies.

Similarly we can represent Red's strategies by a vector (b_J, b_Q, b_K) where $b_i = 0, 1, 2, 3$ gives the instructions for card i.

$$b_i = \begin{cases} 0 & \text{means pass if Blue passes, pass if Blue bets} \\ 1 & \text{means pass if Blue passes, bet if Blue bets} \\ 2 & \text{means bet if Blue passes, pass if Blue bets} \\ 3 & \text{means bet if Blue passes, bet if Blue bets.} \end{cases}$$

Hence Red has $4 \times 4 \times 4 = 64$ strategies.

4. RELATIONS AMONG EXPECTATIONS

Suppose Blue has m strategies and Red has n strategies; then the game is determined by the $m \times n$ matrix $A = (a_{ij})$ where a_{ij} is Blue's payoff if he uses his ith strategy and Red uses his jth strategy. Since we are assuming that the payoff is from Red to Blue, Red's payoff is $-a_{ij}$. Blue's objective in the game is to make a_{ij} as large as possible, whereas Red wants to make $-a_{ij}$ as large as possible, or a_{ij} as small as possible. In terms of this payoff to Blue, we may refer to Blue as the maximizing player and Red as the minimizing player.

Now for any strategy i which Blue may choose, he can be sure of getting at least

$$\min_{j \leq n} a_{ij},$$

where the minimum is taken over all of Red's strategies. Blue is at liberty to choose i; therefore, he can make his choice in such a way as to insure that he gets at least

$$\max_{i \leq m} \min_{j \leq n} a_{ij}.$$

Similarly, for any strategy j which Red may choose, he can be sure of getting at least

$$\min_{i \leq m} (-a_{ij}) = - \max_{i \leq m} a_{ij}.$$

That is, for any strategy j which Red may choose, he can be sure that Blue gets no more than

$$\max_{i \leq m} a_{ij}.$$

Since Red is at liberty to choose j, he can choose it in such a way that Blue will get at most

$$\min_{j \leq n} \max_{i \leq m} a_{ij}.$$

Therefore, there exists a way for Blue to play so that Blue gets at least

$$\max_{i \leq m} \min_{j \leq n} a_{ij},$$

and there exists a way for Red to play so that Blue gets no more than

$$\min_{j \leq n} \max_{i \leq m} a_{ij}.$$

In general, these two quantities are different, but satisfy the relationship

$$\max_{i \leq m} \min_{j \leq n} a_{ij} \leq \min_{j \leq n} \max_{i \leq m} a_{ij}.$$

We prove the latter inequality as follows: Given any i, then

$$\min_{j \leq n} a_{ij} \leq a_{ij} \qquad \text{for all } j.$$

Given any j, then

$$\max_{i \leq m} a_{ij} \geq a_{ij} \qquad \text{for all } i.$$

Hence we have

$$\min_{j \leq n} a_{ij} \leq a_{ij} \leq \max_{i \leq m} a_{ij}$$

or

$$\min_{j \leq n} a_{ij} \leq \max_{i \leq m} a_{ij}.$$

Since the right-hand side of the preceding inequality is independent of i, we have, by taking the maximum of both sides,

$$\max_{i \leq m} \min_{j \leq n} a_{ij} \leq \max_{i \leq m} a_{ij}.$$

Now the left-hand side of the preceding inequality is independent of j. By taking the minimum of both sides, we have

$$\max_{i \leq m} \min_{j \leq n} a_{ij} \leq \min_{j \leq n} \max_{i \leq m} a_{ij}. \tag{1.1}$$

Another way of establishing the inequality is the following: Suppose

$$\max_{i \leq m} \min_{j \leq n} a_{ij} = a_{rs}$$

$$\min_{j \leq n} \max_{i \leq m} a_{ij} = a_{\alpha\beta}.$$

Then it follows that a_{rs} is the minimum of the rth row, or

$$a_{rs} \leq a_{r\beta}.$$

Also, $a_{\alpha\beta}$ is the maximum of the βth column; thus

$$a_{r\beta} \leq a_{\alpha\beta}.$$

Thus we have

$$a_{rs} \le a_{r\beta} \le a_{\alpha\beta}$$

and the inequality is proven.

Examples. (a) Suppose the game is such that its payoff matrix is given by

$$A = \begin{bmatrix} 2 & 3 & 4 & -1 \\ 5 & -2 & 2 & -3 \\ 4 & 1 & 3 & 2 \end{bmatrix}.$$

Then $\max \min a_{ij} = 1$ and $\min \max a_{ij} = 2$. In this game, Blue can receive at least 1. Blue can guarantee this amount by playing his third strategy. The most Red needs to pay or the most that Blue can get is 2. Red can assure this upper bound by playing his fourth strategy.

(b) In Example 2, the game Morra, we have

$$\max \min a_{ij} = -3 \quad \text{and} \quad \min \max a_{ij} = 3.$$

There is a way for Blue to play so that he loses no more than 3 and for Red to play so that Blue wins no more than 3.

5. SADDLE-POINTS

If it happens that the inequality (1.1) becomes an equality, or that

$$\max_{i \le m} \min_{j \le n} a_{ij} = \min_{j \le n} \max_{i \le m} a_{ij} = v \tag{1.2}$$

then Blue can choose a strategy so as to get at least this common value, and Red can keep Blue from getting more than v. In this case, there are strategies i^* and j^* for the two players such that, for all i and j,

$$a_{ij^*} \le a_{i^*j^*} \le a_{i^*j} \tag{1.3}$$

and

$$a_{i^*j^*} = v.$$

Thus Blue cannot do better than to choose i^*; similarly, Red cannot do better than to choose j^*.

We refer to i^*, j^* as *optimal strategies* of Blue and Red, respectively. The optimal strategies have the following properties:

(i) If Blue chooses i^*, then no matter what strategy Red chooses, Blue can get at least v.

(ii) If Red chooses j^*, then no matter what strategy Blue chooses, Blue can get at most v.

(iii) If Blue were to announce in advance that he planned to play strategy i^*, Red could not thereby take advantage of this information and reduce Blue's payoff. Similarly, if Red were to announce j^*, Blue could not increase his payoff.

If the condition (1.2) is satisfied, then the matrix $A = (a_{ij})$ is said to have a *saddle-point* at i^*, j^* and its value is $a_{i^*j^*} = v$. We call v the *value* of the game. It represents the amount that Blue should pay Red at the start of the game in order to equalize the game with respect to winnings.

It is easily seen that conditions (1.2) and (1.3) are equivalent: thus condition (1.2) is satisfied if and only if there exists a pair of strategies i^*, j^* satisfying (1.3). From (1.3) we also have that

$$\max_{i \leq m} a_{ij^*} = \min_{j \leq n} a_{i^*j} = v.$$

Hence a necessary and sufficient condition that a game have a saddle-point is that there exists a member of the payoff matrix which is simultaneously the minimum of its row and the maximum of its column.

A game may have several saddle-points. In such a case all the saddle-points have the same value. Each location of a saddle-point provides another solution or optimal strategy.

Example. The matrix

$$\begin{bmatrix} 1 & -3 & -2 \\ 2 & 5 & 4 \\ 2 & 3 & 2 \end{bmatrix}$$

has a saddle-point in its lower left-hand corner, since this payoff, 2, is simultaneously the minimum of the third row and the maximum of the first column. Hence Blue can expect to get at least 2 in this game (by choosing the third row) and Red can keep Blue from exceeding 2 (by choosing the first column). These strategies, $i^* = 3$, $j^* = 1$, are optimal strategies of Blue and Red, respectively. The value of the game is 2.

Although the lower right-hand corner element of this payoff matrix is 2, which is the value of the game, it is not a saddle-point. This element fails to be the maximum payoff in its column. However, $i^* = 2$, $j^* = 1$ locates another saddle-point.

The pair of optimal strategies, i^*, j^*, is also said to be a solution of the game having a value $v = a_{i^*j^*}$.

Example 7. Armaments.

Suppose that fighter planes can be equipped with any of these armaments—guns, rockets, and toss-bombs—or for ramming. These fighters are to be used against bombers of three types: full firepower and low speed, partial firepower and medium speed, and no firepower and high speed. We wish to determine the best type of armament for the fighter and the best type of bomber. Assume that we are able to establish the effectiveness of each type of fighter versus each type of bomber. This represents the payoff and is measured by a bomber attrition factor per fighter.

Let Blue be the Fighter Command which has four strategies and let

Red be the Bomber Command which has three strategies. A strategy is to be picked by each player independently of the choice of the other. Let the payoff to Blue be the probability that the fighter destroys the bomber. Suppose, for example, the payoff matrix is as follows:

ARMAMENTS PAYOFF

		Bomber Command (Red)		
		1. Full fire, low speed	2. Partial fire, medium speed	3. No fire, high speed
Fighter Command (Blue)	1. Guns	0.30	0.25	0.15
	2. Rockets	0.18	0.14	0.16
	3. Toss-bombs	0.35	0.22	0.17
	4. Ramming	0.21	0.16	0.10

The least that the Fighter Command can get is 0.15, 0.14, 0.17, or 0.10, depending upon whether he uses guns, rockets, toss-bombs, or ramming. These are the minima of the rows. Therefore, Blue will pick toss-bombs, since they yield the most, 0.17. Now the most the Bomber Command will have to pay is 0.35, 0.25, or 0.17, depending on which firepower he adopts. These numbers are the maxima of the columns. Therefore, by picking no firepower, he pays the least, 0.17.

The best strategy for Blue is his third strategy, toss-bombs. The best strategy for Red is also the third, no firepower. The value of the game to the fighter is 0.17. If Blue uses toss-bombs, he is sure of getting at least 0.17. If Red uses no firepower, he is sure of losing no more than 0.17. This is the best that both players can do.

6. GAMES WITH PERFECT INFORMATION

We have seen that if a game has a saddle-point, it is formally easy to find optimal strategies of this game. We look for an element in the payoff matrix which is simultaneously the minimum of its row and the maximum of its column. The location of this element yields optimal strategies of the two players. Having found a saddle-point, we have determined a solution of the game. Of course, the size of the payoff matrix determines the relative ease of finding the saddle-point.

There exists a large class of games, referred to as games with *perfect information*, that have saddle-points. In a game with perfect information, the players move alternately and at each move a player is completely informed about the previous moves, chance or personal, in the game. Examples of games with perfect information are chess and backgammon. Bridge and poker are not games of perfect information since the player does not know the cards dealt to the other players.

Thus, if the strategies for chess were enumerated, we could find optimal strategies by locating the saddle-points in a payoff matrix whose entries consist of $+1, 0, -1$. The existence of a saddle-point in the game of chess follows from the fact that it is a game with perfect information. However, because of the large number of strategies and the size of the payoff matrix, the value of the game (which will be $+1$, 0, or -1) and the optimal strategies for chess have not been computed.

2 THE FUNDAMENTAL THEOREM

1. PRELIMINARIES

We have seen that if the payoff matrix $A = (a_{ij})$ is such that

$$\max_{i \le m} \min_{j \le n} a_{ij} = \min_{j \le n} \max_{i \le m} a_{ij}, \tag{2.1}$$

then Blue has a strategy i^* and Red has a strategy j^* such that

$$\begin{aligned} a_{i^*j} &\geq a_{i^*j^*} \qquad \text{for all } j, \\ a_{ij^*} &\leq a_{i^*j^*} \qquad \text{for all } i. \end{aligned} \tag{2.2}$$

We called i^*, j^* optimal strategies of Blue and Red, respectively, and the pair i^*, j^* is said to be a solution of the game. In this case the solution can be obtained by finding the element of the matrix (a_{ij}) which is simultaneously the minimum of its row and the maximum of its column. Further, even if a player announces, prior to the play of a game, the optimal strategy he plans to use, his opponent cannot take advantage of this information. In other words, if (2.1) holds there is no reason for secrecy with respect to strategy choices on the part of either player.

Example. In the game whose payoff matrix is

$$\begin{bmatrix} 1 & 4 & 1 \\ 2 & 3 & 4 \\ 0 & -2 & 7 \end{bmatrix},$$

there is a saddle-point at $i = 2$, $j = 1$ and its value is 2. If Blue announces he plans to play strategy $i = 2$, then Red cannot reduce Blue's winnings below 2, since Red will then do best by choosing $j = 1$. Similarly, if Red announces he plans to play his first strategy, Blue cannot increase his winnings above 2, since Blue will then choose $i = 2$.

2. GAMES WITHOUT SADDLE-POINTS

Let us now consider games whose payoff matrices are such that

$$\max_{i \leq m} \min_{j \leq n} a_{ij} < \min_{j \leq n} \max_{i \leq m} a_{ij}. \tag{2.3}$$

The left-hand side of the inequality represents the least Blue can get and the right-hand side represents the most Blue can get. In other words, Blue has a strategy which assures him of getting at least

$$\max_{i \leq m} \min_{j \leq n} a_{ij}$$

and Red has a strategy which assures him that Blue cannot get more than

$$\min_{j \leq n} \max_{i \leq m} a_{ij}.$$

Since these two quantities are unequal, we have not arrived at a solution to the game.

Example. The game defined by the payoff matrix

$$\begin{bmatrix} 5 & 1 \\ 3 & 4 \end{bmatrix}$$

does not have a saddle-point. For here we have

$$\min_{j \leq 2} \max_{i \leq 2} a_{ij} = 4,$$

$$\max_{i \leq 2} \min_{j \leq 2} a_{ij} = 3,$$

and

$$\max \min a_{ij} < \min \max a_{ij}.$$

Blue can guarantee himself a payoff of 3 units. However, Red can guarantee that it won't cost him more than 4 units. Therefore, Blue will try to get more than 3 units and Red will try to cut his cost below 4 units, or Blue should win between 3 and 4 per play.

Let us examine Blue's reasoning in this situation. Suppose Red could discover Blue's optimal strategy, perhaps by computing it. Then if this strategy is Blue's first strategy, Red can drive Blue's winnings down to 1; if the strategy is Blue's second strategy, his winnings will be 3. Therefore, if Blue's strategy were discovered, his winnings would be either 1 or 3 on each play. However, Blue is trying to get between 3 and 4. Thus, Blue is at a disadvantage if Red knows Blue's strategy.

It seems then that Blue should concentrate on trying to prevent Red from discovering the strategy. One way would be for Blue to choose his strategy at random. Red then could not discover Blue's strategy because Blue would not know the strategy in advance of the play. On the other hand, if Blue knew Red's strategy then Blue could get either 5 or 4, whereas Red is trying for between 3 and 4 per play. Hence Red should also choose his strategy at random.

3. MIXED STRATEGIES

In a game without a saddle-point, we saw that a player's choice of a strategy will depend on his opponent's choice. Therefore, it is very important that a player find out his opponent's choice of strategy—hence, it is essential that a player's choice of a strategy be unknown to his opponent. Each player will concentrate on keeping his own intentions secret. One way to do this is by using a random device for selecting a strategy.

We shall now extend the notion of a strategy which will satisfy the preceding discussion. A player, instead of choosing a single strategy, may leave the choice of the strategy to chance. That is, he may choose a probability distribution over his set of strategies and then the associated random device selects the particular strategy for the play of the game. Such a probability distribution over the whole set of strategies of a player is a *mixed strategy*.

The function of a mixed strategy is to keep the opponent from discovering the strategy. If a game has a saddle-point, then there is no disadvantage to a player if his chosen strategy is found out by his opponent. However, if the game does not have a saddle-point, it is a definite disadvantage for a player to have his strategy discovered by his opponent. A mixed strategy provides a method for a player to protect himself against having his particular strategy found out by his opponent; choosing from several different strategies at random, with only their probabilities determined, is an effective protection against being found out. The opponent cannot learn the particular strategy in advance, since the player does not know it himself. The strategy is selected at the last moment with the help of a random device.

The game now requires each player to select independently a mixed strategy. The outcome will now be measured in terms of expectation.

We shall represent mixed strategies by column matrices. Let x_i be the probability of selecting strategy i. Then a mixed strategy, or probability distribution X, for Blue may be represented by a column matrix

$$X = \begin{bmatrix} x_1 \\ x_2 \\ \cdot \\ \cdot \\ \cdot \\ x_m \end{bmatrix},$$

where $$x_i \geq 0, \qquad i = 1, 2, \ldots, m$$

and $$\sum_{i=1}^{m} x_i = 1.$$

Similarly, if y_j is the probability of selecting strategy j, then a mixed strategy for Red is a column matrix

$$Y = \begin{bmatrix} y_1 \\ y_2 \\ \cdot \\ \cdot \\ \cdot \\ y_n \end{bmatrix},$$

where $$y_j \geq 0, \qquad j = 1, 2, \ldots, n$$

and $$\sum_{j=1}^{n} y_j = 1.$$

If it happens that $x_i = 1$ for some i, then X is called a *pure* strategy.

Having defined mixed strategies as probability distributions, we need to compute expected payoffs. Suppose Blue chooses strategy i and Red chooses mixed strategy Y; the expected payoff to Blue is

$$h_i = \sum_{j=1}^{n} a_{ij} y_j,$$

which is given by the ith component of the column matrix

$$H = AY = \begin{bmatrix} h_1 \\ h_2 \\ \cdot \\ \cdot \\ \cdot \\ h_m \end{bmatrix}.$$

If Red uses strategy j and Blue uses mixed strategy X, the expected payoff to Blue is

$$k_j = \sum_{i=1}^{m} a_{ij} x_i,$$

which is the jth component of the row matrix K', where

$$K' = X'A = (k_1, k_2, \ldots, k_n).$$

If Blue and Red use mixed strategies, X, Y, respectively, then the expected payoff to Blue is

$$E = X'AY = \sum_{j=1}^{n} \sum_{i=1}^{m} a_{ij} x_i y_j = K'Y = X'H.$$

Example. In the Colonel Blotto payoff matrix (Example 4), if the enemy uses the mixed strategy

$$Y = \begin{bmatrix} \frac{1}{4} \\ 0 \\ \frac{1}{2} \\ \frac{1}{4} \end{bmatrix}$$

and Colonel Blotto uses a pure strategy, Blotto's expectation for each of his pure strategies is $H = AY$, or

$$H = \begin{bmatrix} 4 & 0 & 2 & 1 \\ 0 & 4 & 1 & 2 \\ 1 & -1 & 3 & 0 \\ -1 & 1 & 0 & 3 \\ -2 & -2 & 2 & 2 \end{bmatrix} \begin{bmatrix} \frac{1}{4} \\ 0 \\ \frac{1}{2} \\ \frac{1}{4} \end{bmatrix} = \begin{bmatrix} 2\frac{1}{4} \\ 1 \\ 1\frac{3}{4} \\ \frac{1}{2} \\ 1 \end{bmatrix},$$

where the components of H represent Colonel Blotto's receipts corresponding to each one of his five pure strategies. Now, if the two players' strategies are

$$X = \begin{bmatrix} \frac{1}{4} \\ 0 \\ 0 \\ \frac{1}{4} \\ \frac{1}{2} \end{bmatrix}, \quad Y = \begin{bmatrix} \frac{1}{4} \\ 0 \\ \frac{1}{2} \\ \frac{1}{4} \end{bmatrix},$$

then Blotto's expectation is $E = X'AY = X'H$, or

$$[\tfrac{1}{4}, 0, 0, \tfrac{1}{4}, \tfrac{1}{2}] \begin{bmatrix} 2\frac{1}{4} \\ 1 \\ 1\frac{3}{4} \\ \frac{1}{2} \\ 1 \end{bmatrix} = \tfrac{19}{16}.$$

4. GRAPHICAL REPRESENTATION OF MIXED STRATEGIES

It is possible to represent graphically the expectation of a player as a function of his mixed strategies. If a player has two strategies, the graphical representation is two-dimensional; for three strategies, the representation requires three dimensions. For example, suppose we have the following payoff matrix:

		Red strategies	
		[1]	[2]
Blue strategies	①	5	1
	②	3	4

We can represent the various mixtures of Blue strategies ① and ② by points on the line S_1S_2 of unit length (Fig. 2). Thus Q represents $\frac{3}{4}$ of ① and $\frac{1}{4}$ of ②. If Red uses his strategy [1], then Blue's expectations are given by ordinates PQ of the points P of AB corresponding to each of Blue's mixed strategies. Thus $PQ = 4\frac{1}{2}$ is the expectation of Blue if he uses a mixed strategy of $\frac{3}{4}$ of ① and $\frac{1}{4}$ of ② and Red uses strategy [1].

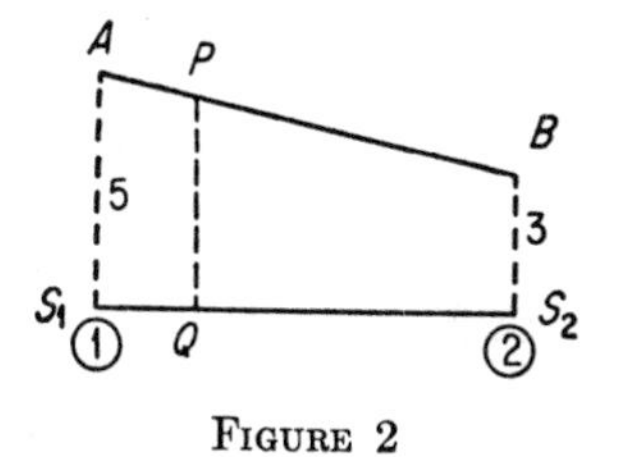

FIGURE 2

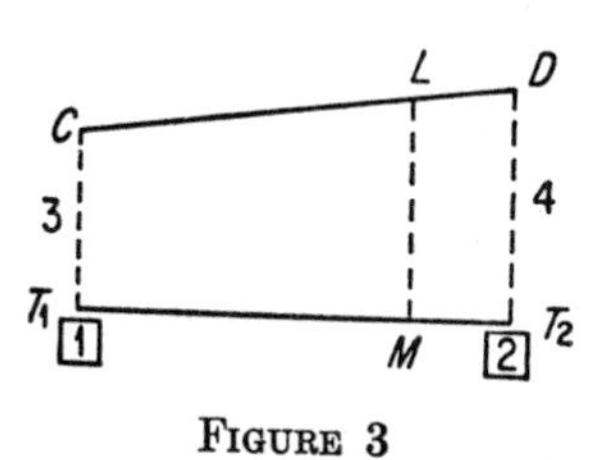

FIGURE 3

If Blue uses strategy ②, then the expectation of Blue for each of Red's mixtures of [1] and [2] is given (Fig. 3) by the ordinates of CD. Thus $LM = 3\frac{4}{5}$ is the expectation of Blue if he uses strategy ② and Red uses a mixture M consisting of $\frac{1}{5}$ of [1] and $\frac{4}{5}$ of [2].

5. THE MINIMAX THEOREM

Having introduced mixed strategies, we now interpret a game with payoff matrix $A = (a_{ij})$ as consisting of Blue's choosing a mixed strategy X and Red's choosing a mixed strategy Y. The payoff to Blue is the expected payoff given by

$$E = X'AY = \sum_{j=1}^{n} \sum_{i=1}^{m} a_{ij}x_iy_j. \tag{2.4}$$

Although a player chooses the mixed strategy, the particular strategy which eventually plays the game is chosen by chance, subject to the given probability distribution.

Suppose Blue chooses his strategy by using a mixed strategy X. Then he can expect to receive at least

$$\min_{Y} X'AY,$$

where the minimum is taken over all possible mixed strategies available to Red. Now since Blue has the choice of X, he will select X so that this minimum is as large as possible. Hence Blue can pick a mixed strategy, call it X^*, which will guarantee him an expectation of at least

$$\max_{X} \min_{Y} X'AY,$$

irrespective of what Red does. Similarly, for each mixed strategy, Y, chosen by Red, the most he will have to pay to Blue is

$$\max_{X} X'AY,$$

where the maximum is taken over all mixed strategies available to Blue. Now Red can choose Y so that the latter quantity is as small as possible. Hence, Red can pick a mixed strategy, Y^*, which will make the expectation of Blue at most

$$\min_Y \max_X X'AY,$$

irrespective of what Blue does. It is apparent from the above remarks, and it can be easily shown, that

$$\max_X \min_Y X'AY \leq \min_Y \max_X X'AY. \tag{2.5}$$

The *minimax theorem* states that these two quantities always have a common value, v, or that

$$\max_X \min_Y X'AY = \min_Y \max_X X'AY = v. \tag{2.6}$$

This remarkable result is the fundamental theorem of game theory. We shall prove this theorem in section 8.

6. OPTIMAL MIXED STRATEGIES

From (2.6) it follows that Blue has a mixed strategy X^* and Red has a mixed strategy Y^* such that

$$\begin{aligned} X^{*\prime}AY &\geq v \qquad \text{for all } Y, \\ X'AY^* &\leq v \qquad \text{for all } X, \\ X^{*\prime}AY^* &= v. \end{aligned} \tag{2.7}$$

The pair X^*, Y^* is called a *solution* of the game, and v, the *value* of the game. We also refer to X^*, Y^* as *optimal* strategies since

(i) If Blue uses X^*, his expectation is at least v, irrespective of what Red does.

(ii) If Red uses Y^*, he can make Blue's expectation at most v, irrespective of what Blue does.

Since $X^{*\prime}AY \geq v$ for all Y and $X'AY^* \leq v$ for all X, the solution also has the property that a player may announce, in advance of the play, the mixed strategy to be employed, and the opponent will be unable to reduce the expectation of the player as a result of this extra knowledge. Of course, this announcement gives information only about the mixed strategy to be employed. It does not give any information about the pure strategy which will actually be used in playing the game. Neither player knows the particular pure strategy, since that is left to a chance device.

Thus, if Blue uses an optimal mixed strategy X^* and Red knows this in advance, Red need not shift from an optimal strategy Y^* as a result of the information. A similar argument applies to Blue.

Example. In Example 4, the Colonel Blotto game, it can be verified by (2.7) that an optimal mixed strategy X^* of Colonel Blotto and Y^* of the enemy is given by

$$X^* = \begin{bmatrix} \frac{4}{9} \\ \frac{4}{9} \\ 0 \\ 0 \\ \frac{1}{9} \end{bmatrix}, \qquad Y^* = \begin{bmatrix} \frac{1}{18} \\ \frac{1}{18} \\ \frac{4}{9} \\ \frac{4}{9} \end{bmatrix}.$$

Colonel Blotto should pick strategies 1, 2, 5 at random with probabilities $\frac{4}{9}$, $\frac{4}{9}$, $\frac{1}{9}$, respectively. The enemy should pick strategies 1, 2, 3, 4 at random with probabilities $\frac{1}{18}$, $\frac{1}{18}$, $\frac{4}{9}$, $\frac{4}{9}$, respectively. The value of the game is $\frac{14}{9}$ to Blotto; i.e., Colonel Blotto, by playing optimally, can expect to receive at least $\frac{14}{9}$ regiments from the enemy. If Colonel Blotto uses his optimal mixed strategy, then, no matter what the enemy does, Colonel Blotto receives at least $\frac{14}{9}$ regiments. If the enemy uses his optimal mixed strategy, then Colonel Blotto may play any strategy except his third and fourth and still receive $\frac{14}{9}$ regiments.

7. GRAPHICAL REPRESENTATION OF MINIMAX THEOREM

If one player has two strategies and the other has any number of strategies, it is possible to illustrate graphically, in two dimensions, the minimax theorem and also solve the game. For example, if the payoff matrix is

		Red strategies	
		[1]	[2]
Blue	①	5	1
strategies	②	3	4

the graph shown in Fig. 4 summarizes the expectation of Blue for various mixed strategies of Blue and Red. The broken line AB represents Blue's expectation for each of his mixed strategies if Red uses pure strategy [1].

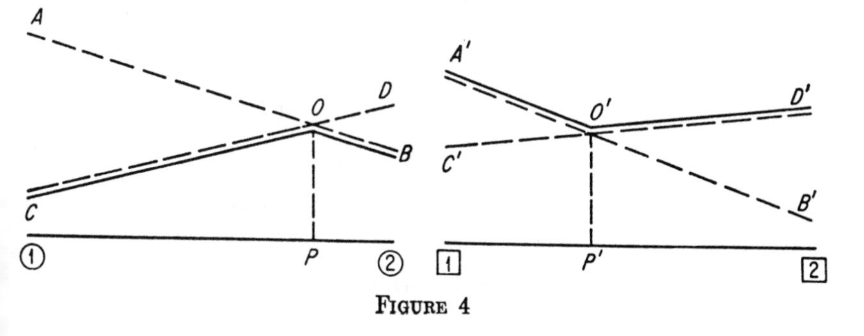

FIGURE 4

The broken line CD represents Blue's expectation for each of his mixed strategies if Red uses strategy [2]. The line $A'B'$ represents the payments by Red for each of his mixed strategies, if Blue uses pure strategy ①. The line $C'D'$ represents the payments by Red if Blue uses pure strategy ②.

The solid lines COB, therefore, represent the least that Blue can receive for each of Red's mixed strategies. It is apparent that Blue will pick the maximum point of COB and Red will pick the minimum point of $A'O'D'$. From the minimax theorem these two points are equal, and the value of the game is $OP = O'P' = 3.4$. The points P and P' give the solution of the game and correspond to

$$X^* = \begin{bmatrix} \frac{1}{5} \\ \frac{4}{5} \end{bmatrix}, \qquad Y^* = \begin{bmatrix} \frac{3}{5} \\ \frac{2}{5} \end{bmatrix}.$$

8. PROOF OF THE MINIMAX THEOREM

In this section we shall give a proof of the minimax theorem, the fundamental theorem of games of strategy. We shall prove that for any matrix $A = (a_{ij})$, we have that

$$\max_X \min_Y \sum_{j=1}^{n} \sum_{i=1}^{m} a_{ij}x_iy_j = \min_Y \max_X \sum_{j=1}^{n} \sum_{i=1}^{m} a_{ij}x_iy_j = v, \tag{2.8}$$

where X and Y are the sets of probability distributions over Blue's and Red's strategies, respectively. From this theorem it follows that every finite zero-sum, two-person game has optimal mixed strategies.

There exist many proofs of this fundamental theorem. Some of the proofs use fixed-point theorems; others are based on a separation theorem of convex sets. Almost all these proofs are existence proofs—they prove the existence of optimal strategies but are not constructive from the viewpoint of practical computation.

We shall give an elementary proof of the minimax theorem which is purely algebraic, requiring nothing more advanced than the notion of an inverse of a matrix. At the same time, the proof will provide an efficient method for computing optimal mixed strategies for both players.

First, let us derive a relationship that exists between the maximum over mixed strategies and the maximum over pure strategies. Suppose $\phi_1, \phi_2, \ldots, \phi_m$ are m arbitrary numbers, and let $X = (x_1, x_2, \ldots, x_m)$ be a mixed strategy of m pure strategies. Then for all i

$$\phi_i \leq \max_i \phi_i = \phi_k,$$

where the max is taken over the m strategies. Multiplying by x_i and summing, we get

$$\sum_{i=1}^{m} \phi_ix_i \leq \max_i \phi_i = \phi_k$$

or that

$$\max_X \sum_{i=1}^{m} \phi_i x_i \leq \max \phi_i = \phi_k.$$

Using the particular mixed strategy which assigns a probability of 1 to the kth strategy, we have

$$\max_X \sum_{i=1}^{n} \phi_i x_i \geq \phi_1 \cdot 0 + \phi_2 \cdot 0 + \ldots + \phi_k \cdot 1 + \ldots + \phi_n \cdot 0 = \phi_k.$$

Hence it follows that

$$\max_X \sum_{i=1}^{m} \phi_i x_i = \max_i \phi_i. \tag{2.9}$$

Similarly, we have

$$\min_Y \sum_{j=1}^{n} \phi_j y_j = \min_j \phi_j. \tag{2.10}$$

Equivalent problem. Let $l(X)$ and $u(Y)$ be defined as follows:

$$l(X) = \min_Y \sum_{j=1}^{n} \sum_{i=1}^{m} a_{ij} x_i y_j = \min_j \sum_{i=1}^{m} a_{ij} x_i,$$

$$u(Y) = \max_X \sum_{j=1}^{n} \sum_{i=1}^{m} a_{ij} x_i y_j = \max_i \sum_{j=1}^{n} a_{ij} y_j.$$

Then, from the definition of a maximum and minimum, we have the following two inequalities:

$$\sum_{i=1}^{m} a_{ij} x_i \geq l(X) \qquad \text{for all } j,$$

$$\sum_{j=1}^{n} a_{ij} y_j \leq u(Y) \qquad \text{for all } i.$$

Now multiplying the first inequality by y_j and summing over j, and multiplying the second inequality by x_i and summing over i, we get

$$l(X) \leq u(Y).$$

In terms of these variables, the minimax theorem states that

$$\max_X l(X) = \min_Y u(Y).$$

Now suppose we can exhibit a pair of mixed strategies, X^* and Y^*, with the property that

$$l(X^*) = u(Y^*).$$

Then since

$$l(X^*) \leq \max_X l(X) \leq u(Y^*),$$

it follows that

$$l(X^*) = \max_X l(X).$$

Similarly,

$$u(Y^*) = \min_{Y} u(Y).$$

Therefore the minimax theorem is essentially a statement of the following equivalent problem: Find an m-component vector X^* and an n-component vector Y^* whose components x_i and y_j, respectively, satisfy the following seven conditions:

$$\begin{aligned}
&\text{I.} \quad x_i \geq 0, && i = 1, 2, \ldots, m;\\
&\text{II.} \quad \sum_{i=1}^{m} x_i = 1;\\
&\text{III.} \quad l(X^*) \leq \sum_{i=1}^{m} a_{ij}x_i, && j = 1, 2, \ldots, n;\\
(2.11)\quad &\text{IV.} \quad y_j \geq 0, && j = 1, 2, \ldots, n;\\
&\text{V.} \quad \sum_{j=1}^{n} y_j = 1;\\
&\text{VI.} \quad \sum_{j=1}^{n} a_{ij}y_j \leq u(Y^*), && i = 1, 2, \ldots, m;\\
&\text{VII.} \quad l(X^*) = u(Y^*).
\end{aligned}$$

Our proof consists of constructing an X^* and Y^* satisfying the seven relations above.

Basis. We begin the proof by defining the augmented matrix G of the game matrix $A = (a_{ij})$ as follows:

$$(2.12)\quad G = \begin{array}{c} \begin{matrix} P_0 & P_1 & P_2 & \ldots & P_n & P_{n+1} & P_{n+2} & \ldots & P_{n+m} \end{matrix} \\ \begin{bmatrix} 0 & 1 & 1 & \ldots & 1 & 0 & 0 & \ldots & 0 \\ -1 & a_{11} & a_{12} & \ldots & a_{1n} & 1 & 0 & \ldots & 0 \\ -1 & a_{21} & a_{22} & \ldots & a_{2n} & 0 & 1 & \ldots & \cdot \\ \cdot & \cdot & \cdot & \ldots & \cdot & \cdot & \cdot & \ldots & \cdot \\ \cdot & \cdot & \cdot & \ldots & \cdot & \cdot & \cdot & \ldots & \cdot \\ \cdot & \cdot & \cdot & \ldots & \cdot & \cdot & \cdot & \ldots & \cdot \\ -1 & a_{m1} & a_{m2} & \ldots & a_{mn} & 0 & 0 & \ldots & 1 \end{bmatrix} \end{array}.$$

This matrix has $m + 1$ rows and $n + m + 1$ columns. Let us label the initial row of G as $i = 0$. Let us label the columns from left to right as follows:

$$P_0; P_1, \ldots, P_n; P_{n+1} = U_1, U_2, \ldots, U_m = P_{n+m},$$

where U_i are vectors with 1 as the ith component and zeros elsewhere. Further, let us assume that we have originally arranged the rows of A such that

$$(2.13)\qquad a_{m1} = \max_{i<m} a_{i1}.$$

Suppose we choose $m + 1$ columns of G. They determine an $m + 1$ square matrix B. We shall call B a *basis* if B satisfies the following conditions:

B_1. P_0 is included as the initial column of B;
B_2. B is nonsingular—i.e., B has an inverse, B^{-1};
B_3. Each row of B^{-1}, except possibly the initial row which we label as $i = 0$, has its first nonzero component positive.

An example of a basis is the following matrix:

$$(2.14) \quad B^0 = (P_0, P_1, U_1, \ldots, U_{m-1}) = \begin{bmatrix} 0 & 1 & 0 & 0 & \ldots & 0 \\ -1 & a_{11} & 1 & 0 & \ldots & 0 \\ -1 & a_{21} & 0 & 1 & \ldots & 0 \\ . & . & . & 0 & \ldots & . \\ . & . & . & . & \ldots & . \\ . & . & . & . & \ldots & . \\ . & . & . & . & \ldots & . \\ . & . & . & . & \ldots & 1 \\ -1 & a_{m1} & 0 & 0 & \ldots & 0 \end{bmatrix}.$$

It is readily verified that B^0 is nonsingular and its inverse is given by

$$(2.15) \qquad (B^0)^{-1} = \begin{bmatrix} a_{m1} & 0 & \ldots & 0 & -1 \\ 1 & 0 & \ldots & 0 & 0 \\ b_1 & 1 & \ldots & 0 & -1 \\ b_2 & 0 & \ldots & 0 & -1 \\ . & . & \ldots & . & . \\ . & . & \ldots & . & . \\ . & . & \ldots & . & . \\ . & . & \ldots & 0 & . \\ b_{m-1} & 0 & \ldots & 1 & -1 \end{bmatrix} = \begin{bmatrix} R_0^0 \\ R_1^0 \\ . \\ . \\ . \\ R_m^0 \end{bmatrix},$$

where $b_i = a_{m1} - a_{i1}$. We can also express the inverse matrix in terms of the row vectors R_i^0. From (2.13) it follows that $b_i \geq 0$; hence every row of $(B^0)^{-1}$, with the possible exception of $i = 0$, has its first nonzero component positive.

Ordering of vectors. Condition B_3 in the definition of a basis can be expressed as a comparison of vectors with the null vector. Later in the proof we shall find it convenient to compare two arbitrary vectors—i.e., given two vectors which are different, we wish a rule for selecting, say, the larger vector. We do this by defining a suitable ordering relationship among vectors.

First, let us define equality of vectors. Two vectors $A = (a_i)$ and $B = (b_i)$ are equal if and only if $a_i - b_i = 0$ for every i. That is, A is equal to B (written $A = B$) if and only if the vector $(A - B)$ is the null vector, 0. Now suppose the two vectors A and B are not equal, then the vector

$(A - B)$ contains at least one nonzero component. If the first nonzero component of $(A - B)$ is positive, then we shall say A is greater than B, and write $A \succ B$. If the first nonzero component of $(A - B)$ is negative, then A is smaller than B and we write $A \prec B$.

If the rows of the inverse of a basis B are denoted by $R_0, R_1, \ldots, R_m$, then the condition B_3 defining a basis states

$$R_i \succ 0, \qquad i = 1, 2, \ldots, m. \tag{2.16}$$

From the above method of ordering it follows that the minimum vector of a series of vectors is that vector whose first component is least; if there is a tie, then one compares the second components of the tying vectors, etc.

Optimal basis. Let

$$B = (P_0, P_{j_1}, P_{j_2}, \ldots, P_{j_m})$$

be a basis whose inverse is

$$B^{-1} = \begin{bmatrix} R_0 \\ R_1 \\ \cdot \\ \cdot \\ \cdot \\ R_m \end{bmatrix} = (C_0, C_1, \ldots, C_m),$$

where the R's are row vectors and the C's are column vectors. Then from the fact that

$$B^{-1}B = I,$$

where I is the unit matrix, it follows that

$$R_kP_{j_i} = 0 \qquad \text{for } i \neq k,$$

$$R_iP_{j_i} = 1, \qquad i = k = 0, 1, 2, \ldots, m,$$

where $j_0 = 0$.

In particular, we have

$$R_0P_{j_i} = 0 \qquad \text{for } j_i = j_1, j_2, \ldots, j_m. \tag{2.17}$$

Now suppose we form the $n + m$ scalar products

$$R_0P_j, \qquad j = 1, 2, \ldots, n + m.$$

From (2.17) it follows that at least m of these products will be zero. If the remaining n scalar products are no larger than zero, then we shall call B an *optimal basis*. As will be shown, an optimal basis yields optimal mixed strategies of the players. In other words, if the basis B is such that

$$R_0P_j \leq 0, \qquad j = 1, 2, \ldots, n + m,$$

then B is an optimal basis.

Optimal strategies. Let B be an optimal basis. Let the components of the 0-row and 0-column of B^{-1} be denoted as follows:

(2.18) $$R_0 = (l, -x_1, -x_2, \ldots, -x_m),$$

(2.19) $$C_0 = (u, y_{j_1}, y_{j_2}, \ldots, y_{j_m}).$$

We shall show that an optimal mixed strategy X^* for Blue is obtained from the row R_0 by setting

$$X^* = (x_1, x_2, \ldots, x_m).$$

An optimal mixed strategy $Y^* = (y_1, y_2, \ldots, y_n)$ for Red is obtained from the column C_0 by setting

$$\begin{aligned} y_j &= y_{j_i} && \text{for } j_i \leq n, \\ &= 0 && \text{for all other } j. \end{aligned}$$

We shall prove that X^* and Y^* are optimal mixed strategies of the original game by showing that X^* and Y^* as defined above satisfy the seven conditions I–VII of (2.11).

Suppose we have an optimal basis; let us form the products R_0P_j for different values of j.

If $j = 0$, we have

$$1 = R_0P_0 = \sum_{i=1}^{m} x_i,$$

hence condition II is satisfied.

If $1 \leq j \leq n$, we have

$$0 \geq R_0P_j = l - \sum_{i=1}^{m} a_{ij}x_i,$$

which satisfies condition III.

If $n + 1 \leq j \leq n + m$, we have

$$0 \geq R_0P_j = R_0U_i = -x_i \qquad \text{for } 1 \leq i \leq m,$$

hence condition I is satisfied.

Now since B is a basis, we have

$$R_i \succ 0, \qquad i = 1, 2, \ldots, m.$$

In particular the first component of each R_i is non-negative. But from (2.19) the first component of each R_i is y_{j_i}. Hence IV is satisfied.

Since $BB^{-1} = I$, we have in particular that

$$BC_0 = U_0.$$

From this matrix equation we get $m + 1$ linear equations in the $m + 1$ variables $(u, y_{j_1}, y_{j_2}, \ldots, y_{j_m})$. The first of these equations is

$$\sum_{j_i \leq n} y_{j_i} = \sum_{j=1}^{n} y_j = 1,$$

since the first component of each P_{j_i} is 1 for $1 \leq j_i \leq n$ and 0 otherwise. Hence V is satisfied. Now the remaining m equations may be written as follows:

$$-u + \sum_{j_k \leq n} a_{ij_k} y_{j_k} + \sum_{j_k > n} \delta_{j_k} y_{j_k} = 0, \qquad i = 1, 2, \ldots, m,$$

where $\delta_{j_k} \geq 0$. It follows that

$$\sum_{j=1}^{n} a_{ij} y_j \leq u, \qquad i = 1, 2, \ldots, m$$

and VI is satisfied. Finally VII is satisfied since u and l are both defined by the same element of B^{-1}.

Constructing an optimal basis. The problem of finding optimal strategies has now been reduced to the problem of constructing an optimal basis. The inverse of such a basis will yield optimal strategies. We shall now describe an iterative procedure for constructing an optimal basis and its inverse.

The iterative procedure starts with some basis, for example B^0 as defined in (2.14). If B^0 is not optimal then we construct from B^0 a new basis, B^1, which differs from B^0 by only one column. Further, if R_0^1 is the 0th row of $(B^1)^{-1}$, the inverse of B^1, then R_0^1 will have the property that

$$R_0^0 > R_0^1. \tag{2.20}$$

If B^1 is not optimal, then we iterate the preceding algorithm for B^1, etc.

This process generates a sequence of bases. From (2.20) it follows that no basis can be repeated. Further, the number of bases cannot exceed the number of ways of choosing m columns out of $n + m$ columns of the augmented matrix G. However, we shall terminate the process when we have arrived at an optimal basis, which is always possible, as will be shown.

Suppose the basis B^0 is not optimal, then there exists some P_j in G such that $R_0^0 P_j > 0$. To construct B^1 from B^0 we let P_s replace a column of B^0, where P_s is determined by the condition

$$R_0^0 P_s = \max_j R_0^0 P_j > 0, \qquad j = 1, 2, \ldots, n + m. \tag{2.21}$$

In case the choice of P_s is not unique, then choose P_s with the smallest index.

Next compute the column vector $V = (v_0, v_1, \ldots, v_m)$ satisfying the equation $B^0 V = P_s$. Hence we have

$$v_i = R_i^0 P_s, \qquad i = 0, 1, \ldots, m. \tag{2.22}$$

In particular, we have

$$v_0 = R_0^0 P_s > 0.$$

Further, $v_i > 0$ for some other $i \neq 0$. For, if we assume the contrary, namely that $v_i \leq 0$ $(i \neq 0)$, then from the relationship

$$P_s = B^0 V = v_0 P_0 + \sum_{i=1}^{m} v_i P_{j_i},$$

it follows that

$$P_0 = \frac{1}{v_0} P_s - \sum_{i=1}^{m} \frac{v_i}{v_0} P_{j_i}.$$

Hence the column P_0 can be written as a positive linear combination of $m + 1$ columns of G. From the definition of G, it is clear P_0 cannot be written as a positive linear combination of the other columns, and so we have a contradiction.

We now choose to drop from B^0 that column P_{j_r} such that

$$\min_i \frac{R_i^0}{v_i} = \frac{R_r^0}{v_r} \qquad \text{for } v_i > 0,\, v_r > 0,\, i \neq 0. \tag{2.23}$$

From the above it follows that

$$R_i^0 - \frac{v_i}{v_r} R_r^0 > 0 \qquad \text{for } v_i > 0. \tag{2.24}$$

We have constructed B^1 from B^0 by dropping from B^0 column P_{j_r} and adding column P_s, as defined above. We need to show that B^1 is a basis. We shall do this by constructing $[B^1]^{-1}$ from $[B^0]^{-1}$.

Let the rows of $[B^1]^{-1}$ be designated by R_i^1 for $i = 0, 1, 2, \ldots, m$. We shall now verify that we can get R_i^1 from R_i^0 and v_i by letting

$$\begin{aligned} R_i^1 &= R_i^0 - \frac{v_i}{v_r} R_r^0 \qquad &\text{for } i \neq r, \\ R_r^1 &= \frac{1}{v_r} R_r^0. \end{aligned} \tag{2.25}$$

Using $[B^1]^{-1}$ as defined by (2.25) we shall verify that $[B^1]^{-1}B^1 = I$. First, if $i \neq r$ and $j_i \neq s$,

$$R_k^1 P_{j_i} = R_k^0 P_{j_i} - \frac{v_k}{v_r} R_r^0 P_{j_i} = 0 \qquad \text{for } i \neq k,$$

$$R_k^1 P_{j_k} = R_k^0 P_{j_k} - \frac{v_k}{v_r} R_r^0 P_{j_k} = 1 - 0 = 1 \quad \text{for } i = k.$$

Second, if $i \neq r$ and $j_i = s$, we have

$$R_i^1 P_s = R_i^0 P_s - \frac{v_i}{v_r} R_r^0 P_s = v_i - \frac{v_i}{v_r} v_r = 0.$$

Thirdly, if $i = r$ and $j_i \neq s$, we have

$$R_r^1 P_{j_i} = \frac{1}{v_r} R_r^0 P_{j_i} = 0.$$

Finally, if $i = r$ and $j_i = s$, we have

$$R_r^1 P_s = \frac{R_r^0}{v_r} P_s = \frac{v_r}{v_r} = 1.$$

Therefore (2.25) yields the inverse of the matrix B^1.

In order to complete the proof that B is a basis, we need to show that $R_i^1 \succ 0$ for $i = 1, 2, \ldots, m$. Now for $i = r$, we have $R_r^0 \succ 0$ and $v_r > 0$. Hence

$$R_r^1 = \frac{1}{v_r} R_r^0 \succ 0.$$

If $i \neq r$, and if $v_i \leq 0$, then

$$R_i^1 = R_i^0 - \frac{v_i}{v_r} R_r^0 \succ 0.$$

If $i \neq r$, and if $v_i > 0$, then from (2.24) we have

$$R_i^1 = R_i^0 - \frac{v_i}{v_r} R_r^0 \succ 0.$$

Therefore B^1 is a basis.

Finally, we need to show that no basis can be repeated in this process. It is sufficient that (2.20) is satisfied, or that $R_0^1 \prec R_0^0$. This follows from the fact that R_r^0 is a row of a nonsingular matrix and hence must possess at least one nonzero component, the first of which must be positive. Since $v_0 > 0$ and $v_r > 0$, we have that

$$R_0^1 = R_0^0 - \frac{v_0}{v_r} R_r^0 \prec R_0^0.$$

Note that if P_{j_r} is a column removed from B^0, then we have

$$R_0^1 P_{j_r} = R_0^0 P_{j_r} - \frac{v_0}{v_r} R_r^0 P_{j_r} = 0 - \frac{v_0}{v_r} < 0.$$

We also have

$$R_0^1 P_j = 0 \qquad \text{for } (j = j_1, j_2, \ldots, s, \ldots, j_m).$$

Thus at least $m + 1$ columns of the $m + n$ columns now have the property $R_0^1 P_j \leq 0$. This compares with at least m columns having this property in the previous stage. Further, on the next iteration this P_{j_r} cannot return to a basis since a candidate s for the basis must satisfy $R_0 P_s > 0$. This completes the verification.

Example 8. A Guessing Game.

Let us apply the computational method to the following game: Blue secretly picks one of three numbers 1, 2, 3. Red then proceeds to guess the number picked by announcing his guess. Each time Red announces his guess, Blue answers "high," "low," or "correct," as the case may be. The game continues until Red has guessed correctly. The payoff to Blue is the number of guesses required by Red to identify the number.

A strategy for Blue is the choice of a number 1, 2, or 3. A strategy for Red may be represented by a triplet $(G; H, L)$ where G is the number guessed the first round, H is the number guessed the second round if Red hears "high," and L is the number guessed the second round if Red hears "low." It is clear that the game can be terminated in two rounds. Thus Red has five strategies,

$$(1; 0, 2),\ (1; 0, 3),\ (2; 1, 3),\ (3; 1, 0),\ (3; 2, 0),$$

where 0 means "not applicable."

The payoff to Blue is the number of guesses required by Red to get a "correct" response from Blue. The payoff matrix is the following:

Guessing Game Payoff

		Red's choices				
		(102)	(103)	(213)	(310)	(320)
Blue's choices	1	1	1	2	2	3
	2	2	3	1	3	2
	3	3	2	2	1	1

The augmented matrix associated with this game is given below:

$$G = \begin{bmatrix} 0 & 1 & 1 & 1 & 1 & 1 & 0 & 0 & 0 \\ -1 & 1 & 1 & 2 & 2 & 3 & 1 & 0 & 0 \\ -1 & 2 & 3 & 1 & 3 & 2 & 0 & 1 & 0 \\ -1 & 3 & 2 & 2 & 1 & 1 & 0 & 0 & 1 \end{bmatrix}.$$

(columns $P_0, P_1, P_2, P_3, P_4, P_5, P_6, P_7, P_8$)

First iteration. The initial basis B^0 consists of (P_0, P_1, P_6, P_7), or

$$B^0 = \begin{bmatrix} 0 & 1 & 0 & 0 \\ -1 & 1 & 1 & 0 \\ -1 & 2 & 0 & 1 \\ -1 & 3 & 0 & 0 \end{bmatrix}.$$

The inverse of B^0, or $(B^0)^{-1}$, is computed from (2.15):

$$(B^0)^{-1} = \begin{bmatrix} 3 & 0 & 0 & -1 \\ 1 & 0 & 0 & 0 \\ 2 & 1 & 0 & -1 \\ 1 & 0 & 1 & -1 \end{bmatrix} = \begin{bmatrix} R_0^0 \\ R_1^0 \\ R_2^0 \\ R_3^0 \end{bmatrix}.$$

Using this inverse, we compute $R_0^0 P_j$ for all $j \neq 0$. We find that

$$\max_{j \neq 0} R_0^0 P_j = R_0^0 P_4 = 2.$$

Therefore P_4 replaces some column in B^0. To determine the column to be dropped, we first compute the vector $V = (v_i)$ where $v_i = R_i^0 P_4$. Substituting, we get

$$v_0 = 2, \quad v_1 = 1, \quad v_2 = 3, \quad v_3 = 3.$$

Next we obtain

$$\min_{\substack{v_i>0\\ i\neq 0}} \frac{R_i^0}{v_i} = \frac{R_3^0}{3},$$

hence P_7 is dropped from the basis.

Second iteration. The next basis B^1 is

$$B^1 = (P_0, P_1, P_6, P_4).$$

To get its inverse we set

$$R_i^1 = R_i^0 - \frac{v_i}{v_3} R_3^0 \qquad \text{for } i \neq 3,$$

$$R_3^1 = \frac{1}{v_3} R_3^0.$$

We then get

$$(B^1)^{-1} = \begin{bmatrix} R_0^1 \\ R_1^1 \\ R_2^1 \\ R_3^1 \end{bmatrix} = \begin{bmatrix} \frac{7}{3} & 0 & -\frac{2}{3} & -\frac{1}{3} \\ \frac{2}{3} & 0 & -\frac{1}{3} & \frac{1}{3} \\ 1 & 1 & -1 & 0 \\ \frac{1}{3} & 0 & \frac{1}{3} & -\frac{1}{3} \end{bmatrix}.$$

After computing $R_0^1P_j$ for all $j \neq 0$ we find that

$$\max_{j\neq 0} R_0^1P_j = R_0^1P_3 = 1.$$

Therefore P_3 replaces some column in B^1. The new vector V can now be computed. Its components are

$$v_0 = 1, \quad v_1 = 1, \quad v_2 = 2, \quad v_3 = 0.$$

Using these values, we find

$$\min_{\substack{v_i>0\\ i\neq 0}} \frac{R_i^1}{v_i} = \frac{R_2^1}{2}.$$

Hence P_6 is dropped from B^1.

Third iteration. We have $B^2 = (P_0, P_1, P_3, P_4)$. The inverse of B^2 will be

$$(B^2)^{-1} = \begin{bmatrix} \frac{11}{6} & -\frac{3}{6} & -\frac{1}{6} & -\frac{2}{6} \\ \frac{1}{6} & -\frac{3}{6} & \frac{1}{6} & \frac{2}{6} \\ \frac{3}{6} & \frac{3}{6} & -\frac{3}{6} & 0 \\ \frac{2}{6} & 0 & \frac{2}{6} & -\frac{2}{6} \end{bmatrix}.$$

Using this inverse, we compute that

$$\max_{j\neq 0} R_0^2P_j = R_0^2P_2 = \tfrac{1}{6},$$

$$v_0 = \tfrac{1}{6}, \quad v_1 = \tfrac{5}{6}, \quad v_2 = -\tfrac{3}{6}, \quad v_3 = \tfrac{4}{6},$$

$$\min_{\substack{v_i>0\\ i\neq 0}} \frac{R_i^2}{v_i} = \frac{R_1^2}{\frac{5}{6}}.$$

Fourth iteration. We have $B^3 = (P_0, P_2, P_3, P_4)$ and

$$(B^3)^{-1} = \begin{bmatrix} \frac{54}{30} & -\frac{12}{30} & -\frac{6}{30} & -\frac{12}{30} \\ \frac{1}{5} & -\frac{3}{5} & \frac{1}{5} & \frac{2}{5} \\ \frac{18}{30} & \frac{6}{30} & -\frac{12}{30} & \frac{6}{30} \\ \frac{6}{30} & \frac{12}{30} & \frac{6}{30} & -\frac{18}{30} \end{bmatrix} = \begin{bmatrix} R_0^3 \\ R_1^3 \\ R_2^3 \\ R_3^3 \end{bmatrix}.$$

We find that

$$R_0^3 P_j \leq 0 \qquad \text{for all } j \neq 0.$$

Therefore a solution to the game has been obtained. The optimal strategies are $X^* = (\frac{2}{5}, \frac{1}{5}, \frac{2}{5})$, obtained from the top row of $(B^3)^{-1}$ and $Y^* = (0, \frac{1}{5}, \frac{3}{5}, \frac{1}{5}, 0)$, obtained from the first column of $(B^3)^{-1}$. The value of the game, obtained from the upper left corner, is $\frac{9}{5}$.

3 PROPERTIES OF OPTIMAL STRATEGIES

1. MANY OPTIMAL STRATEGIES

From the minimax theorem it follows that every finite game has a solution in mixed strategies. In some cases, depending on the payoff matrix, the game may have many solutions. In this chapter we study the properties of this set of solutions. We shall first examine some properties of an optimal strategy and then analyze the properties of the set of optimal strategies.

2. SOME PROPERTIES OF AN OPTIMAL STRATEGY

Suppose $X^* = (x_i^*)$ and $Y^* = (y_j^*)$ are optimal strategies of Blue and Red, respectively. If $x_i^* > 0$, then strategy i may be played, depending on the outcome of the randomization. If $x_i^* = 0$, then strategy i is not played.

Let

$$U^* = AY^* = \begin{bmatrix} u_1^* \\ u_2^* \\ \cdot \\ \cdot \\ \cdot \\ u_m^* \end{bmatrix}, \quad L^* = A'X^* = \begin{bmatrix} l_1^* \\ l_2^* \\ \cdot \\ \cdot \\ \cdot \\ l_n^* \end{bmatrix}.$$

Thus

$$U^* = (u_i^*) = (\sum_{j=1}^{n} a_{ij}y_j^*)$$

is a vector whose components represent Blue's expectation if he uses a pure strategy i and Red uses an optimal strategy Y^*. Similarly, the components of

$$L^* = (l_j^*) = (\sum_{i=1}^{m} a_{ij}x_i^*)$$

represent Blue's expectation if he uses an optimal mixed strategy X^* and Red uses a pure strategy j.

From the minimax theorem we have

$$\max_i \sum_{j=1}^{n} a_{ij}y_j^* = \min_j \sum_{i=1}^{m} a_{ij}x_i^* = v. \tag{3.1}$$

Hence we have

(i) $\max_i u_i^* = \min_j l_j^* = v$. There exists at least one pure strategy for each player which, if used against his opponent's optimal mixed strategy, yields the value of the game.

From (3.1) we also get

(ii) $u_i^* \leq v \leq l_j^*$ for all i and j. Against an opponent's optimal mixed strategy a pure strategy cannot yield a higher expected payoff than his optimal mixed strategy.

Since $u_i^* \leq v$ for all i, let us designate by S_1 the set of Blue strategies for which $u_i^* < v$, and by S_2 the set of Blue strategies for which $u_i^* = v$. Then, by the minimax theorem, we have

$$v = \sum_{i=1}^{m} u_i^* x_i^* = v \sum_{S_2} x_i^* + \sum_{S_1} u_i^* x_i^*$$

or

$$v(1 - \sum_{S_2} x_i^*) = \sum_{S_1} u_i^* x_i^*.$$

Now from the definition of S_1 and S_2, it follows that

$$1 - \sum_{S_2} x_i^* = \sum_{S_1} x_i^*.$$

Hence we get

$$v \sum_{S_1} x_i^* = \sum_{S_1} u_i^* x_i^*$$

or

$$\sum_{S_1} (v - u_i^*) x_i^* = 0. \tag{3.2}$$

Since $v - u_i^* > 0$ for every i in S_1, it follows from (3.2) that $x_i^* = 0$ for every i in S_1. We have shown that

(iii) If $u_i^* < v$, then $x_i^* = 0$. A player's optimal mixed strategy contains no pure strategy which yields less than the value of the game when that pure strategy is used against an opponent's optimal mixed strategy.

Since $x_i^* \geq 0$ for all i, we can express the game value v as follows:

$$v = \sum_{i=1}^{m} u_i^* x_i^* = \sum_{x_i^*>0} u_i^* x_i^*$$
$$= \sum_{\substack{x_i^*>0 \\ u_i^*=v}} u_i^* x_i^* + \sum_{\substack{x_i^*>0 \\ u_i^*<v}} u_i^* x_i^*.$$

Hence

$$v = v \sum_{R_2} x_i^* + \sum_{R_1} u_i^* x_i^*,$$

where R_1 represents the set of Blue strategies for which $x_i^* > 0$ and $u_i^* < v$, and R_2 is the set of Blue strategies for which $x_i^* > 0$ and $u_i^* = v$. But since

$$\sum_{R_1} x_i^* + \sum_{R_2} x_i^* = 1,$$

we get

$$v \sum_{R_1} x_i^* = \sum_{R_1} u_i^* x_i^*.$$

Therefore

$$\sum_{R_1} (v - u_i^*) x_i^* = 0.$$

Since we have assumed $x_i^* > 0$, and $v - u_i^* > 0$ for each i in R_1, it follows that R_1 is vacuous. We have

(iv) If $x_i^* > 0$ then $u_i^* = v$. If a pure strategy is a member of an optimal mixed strategy it yields the value of the game when used against an opponent's optimal mixed strategy.

Example. For the Colonel Blotto game (Example 4) we have the payoff matrix

$$A = \begin{bmatrix} 4 & 0 & 2 & 1 \\ 0 & 4 & 1 & 2 \\ 1 & -1 & 3 & 0 \\ -1 & 1 & 0 & 3 \\ -2 & -2 & 2 & 2 \end{bmatrix},$$

for which the value of the game is $v = \frac{14}{9}$ to Blotto. The following vectors can be readily verified.

$$X^* = \begin{bmatrix} \frac{4}{9} \\ \frac{4}{9} \\ 0 \\ 0 \\ \frac{1}{9} \end{bmatrix}, \quad U^* = \begin{bmatrix} \frac{14}{9} \\ \frac{14}{9} \\ \frac{12}{9} \\ \frac{12}{9} \\ \frac{14}{9} \end{bmatrix}, \quad Y^* = \begin{bmatrix} \frac{1}{18} \\ \frac{1}{18} \\ \frac{4}{9} \\ \frac{4}{9} \end{bmatrix}, \quad L^* = \begin{bmatrix} \frac{14}{9} \\ \frac{14}{9} \\ \frac{14}{9} \\ \frac{14}{9} \end{bmatrix}.$$

Note that $u_3^* = \frac{12}{9} < \frac{14}{9} = v$ and $x_3^* = 0$; also, $u_4^* = \frac{12}{9} < \frac{14}{9} = v$ and $x_4^* = 0$.

3. CONVEX SET OF OPTIMAL STRATEGIES

Suppose a player has more than one optimal strategy. Let X_1^* and X_2^* be two optimal mixed strategies of Blue. If v is the value of the game, then from the property of an optimal strategy

$$X_1^{*\prime}AY \geq v \quad \text{and} \quad X_2^{*\prime}AY \geq v \qquad \text{for all } Y.$$

Let λ be an arbitrary number such that $0 < \lambda < 1$. Then we have

$$\lambda X_1^{*\prime}AY + (1 - \lambda)X_2^{*\prime}AY \geq v \qquad \text{for all } Y.$$

Thus

$$[\lambda X_1^{*\prime} + (1 - \lambda)X_2^{*\prime}]AY \geq v \qquad \text{for all } Y. \tag{3.3}$$

Hence, for every value of λ, the mixed strategy $[\lambda X_1^{*\prime} + (1 - \lambda)X_2^*]$, which is a convex linear combination of X_1^* and X_2^*, is also an optimal strategy for Blue. Since λ was arbitrary, it follows that if a player has more than one optimal strategy, he has an infinite number of optimal strategies. This infinite set of optimal strategies satisfying (3.3) is called a *convex* set. A similar argument may be applied to Red's strategies.

Let T_1 be the set of optimal strategies for Blue and T_2 the set of optimal strategies for Red. Then T_1 and T_2 are convex sets. It follows that every member of T_1 is a convex linear combination of certain points $K(T_1)$ of the set T_1. Thus we may describe T_1 by describing all members of $K(T_1)$. It will turn out that $K(T_1)$ is a finite set.

4. OPERATIONS ON GAMES

There are several operations one may perform on the payoff matrix without altering the set of optimal strategies of a game. These operations are sometimes useful in solving games with a large number of strategies.

Permuting. A permutation of the rows or columns of a payoff matrix permutes the components of the solution and does not alter the value of the game. We can also permute the players by solving the game whose payoff matrix is the negative transpose, $-A'$, of the payoff matrix A.

Addition of and multiplication by constant. If a constant, c, is added to each element of the payoff matrix which has a game value, v, then the value of the new game is $v + c$. It is clear that the sets of solutions are the same for both games.

If every element of A is multiplied by a positive number c, then the new game has a value cv. Again, the set of solutions is unchanged.

5. DOMINATED STRATEGIES

Let us define a *poor strategy* of a player as some pure strategy which appears with zero probability in every optimal mixed strategy of that player. That is, we shall say Blue's kth pure strategy is poor if we have $x_k^* = 0$ in every Blue optimal mixed strategy $X^* = (x_i^*)$.

If T_1 is Blue's set of optimal mixed strategies, then k is a poor strategy if for every member of T_1 we have $x_k^* = 0$. Since a poor strategy never appears with positive probability in any optimal mixed strategy, it will never be played. Knowing the poor strategies, if any, we can reduce the size of the game by the number of poor strategies.

One method of examining a game for poor strategies is to examine the payoff matrix for dominances. Suppose that the payoff matrix $A = (a_{ij})$ is such that

$$a_{ij} > a_{kj} \qquad \text{for } j = 1, 2, \ldots, n, \tag{3.4}$$

or that row i dominates strictly row k. Then we say that strategy k of Blue is dominated by strategy i, or k is a dominated strategy. However, for Red, if column r dominates strictly column s, i.e., if

$$a_{ir} > a_{is} \qquad \text{for } i = 1, 2, \ldots, m, \tag{3.5}$$

then strategy r is dominated by strategy s.

Let us assume that for Blue strategy k is dominated by strategy i, or

$$a_{kj} < a_{ij} \qquad \text{for } j = 1, 2, \ldots, n.$$

Then if $Y^* = (y_j^*)$ is an optimal strategy for Red, we have

$$u_k^* = \sum_{j=1}^{n} a_{kj}y_j^* < \sum_{j=1}^{n} a_{ij}y_j^* = u_i^* \leq v,$$

or

$$u_k^* < v.$$

It follows that for any X^* in T_1,

$$x_k^* = 0.$$

Hence strategy k is a poor strategy. Similarly, we can show that if (3.5) is satisfied, then strategy r is a poor strategy for Red. We have thus shown that

(i) If a strategy is dominated by another strategy, the dominated strategy is a poor strategy.

Suppose that there exist pure strategies k, l, and r and a number $\lambda(0 < \lambda < 1)$ such that

$$a_{kj} < \lambda a_{lj} + (1 - \lambda)a_{rj} \qquad \text{for } j = 1, 2, \ldots, n. \tag{3.6}$$

That is, suppose strategy k is dominated by a convex linear combination of strategies l and r. Then for any Y^* we get

$$u_k^* = \Sigma a_{kj}y_j^* < \lambda \Sigma a_{lj}y_j^* + (1 - \lambda) \Sigma a_{rj}y_j^* = \lambda u_l^* + (1 - \lambda)u_r^*.$$

But

$$u_l^* \leq v \quad \text{and} \quad u_r^* \leq v.$$

Hence

$$u_k^* < \lambda v + (1 - \lambda)v = v.$$

Again, this implies that for any X^* in T_1,

$$x_k^* = 0,$$

or the kth strategy is a poor strategy. We have thus shown that

(ii) If a strategy is dominated by a convex linear combination of other strategies, the dominated strategy is a poor strategy.

Now suppose a convex linear combination of strategies k and l is dominated by a convex linear combination of strategies r and s—i.e., suppose there exist λ and μ where $0 \leq \lambda \leq 1$ and $0 \leq \mu \leq 1$, such that

$$\lambda a_{kj} + (1 - \lambda)a_{lj} < \mu a_{rj} + (1 - \mu)a_{sj} \qquad \text{for } j = 1, 2, \ldots, n. \tag{3.7}$$

Multiplying (3.7) by y_j^* and summing over all j, we get

$$\lambda \ \Sigma \ a_{kj}y_j^* + (1 - \lambda) \ \Sigma \ a_{lj}y_j^* < \mu \ \Sigma \ a_{rj}y_j^* + (1 - \lambda) \ \Sigma \ a_{sj}y_j^*.$$

Hence

$$\lambda u_k^* + (1 - \lambda)u_l^* < \mu u_r^* + (1 - \mu)u_s^* \leq v. \tag{3.8}$$

Now $u_k^* \leq v$ and $u_l^* \leq v$. If $u_k^* = v$ and $u_l^* = v$, then from (3.8) we get the contradiction $v < v$. Hence either $u_k^* < v$ or $u_l^* < v$. Therefore, either $x_k^* = 0$ or $x_l^* = 0$. We have shown that

(iii) If a convex linear combination of strategies is dominated by a convex linear combination of other strategies, then there exists at least one poor strategy among the dominated convex combination of strategies.

Finally, suppose that the payoff matrix A can be decomposed into four submatrices A_1, A_2, A_3, A_4 as follows:

$$A = \begin{bmatrix} A_1 & A_2 \\ A_3 & A_4 \end{bmatrix},$$

where the submatrices A_2 and A_3 have the following properties:

(3.9)
(a) Each column of A_2 dominates strictly some column of A_1.
(b) Each row of A_3 is dominated strictly by some row of A_1.

Let X_1^*, Y_1^* be a solution of the game having A_1 as the payoff matrix. Then it is readily verified that the pair of full vectors $X^* = (X_1^*, 0, \ldots, 0)$, $Y^* = (Y_1^*, 0, \ldots, 0)$ is a solution of the original game A. Thus Blue has optimal mixed strategies which mix only those strategies associated with A_1, and Red has similar optimal strategies. We shall now show that these are the only optimal strategies. If Blue has an optimal mixed strategy which contains a pure strategy k determining A_3, then $x_k^* > 0$. Therefore for any optimal Red strategy Y^*,

$$u_k^* = v.$$

But if $Y^* = (Y_1^*, 0, \ldots, 0)$, it follows from the dominance assumption that

$$u_k^* < v,$$

a contradiction to the optimality assumption. Therefore, in every optimal strategy, $x_k^* = 0$ for all k determining A_3. Thus all Blue's strategies which determine A_3 are poor strategies. Similarly the Red strategies which determine A_2 are poor strategies. It is interesting to note that these conclusions are independent of A_4. We have shown that

(iv) If the payoff matrix A can be decomposed in such a way that (3.9) is satisfied, then all the solutions of A are obtained by solving A_1. The remaining strategies are poor strategies.

6. ALL STRATEGIES ACTIVE

We have seen that for any game the expectation of Blue is at most v, the value of the game, if he uses a pure strategy and Red uses an optimal mixed strategy. Similarly, the least that Red can expect to pay Blue is v if Red uses a pure strategy and Blue uses an optimal mixed strategy. Further, each player has at least one pure strategy, which, if used against an opponent's optimal mixed strategy, yields exactly v to Blue. Now it may happen that the game matrix is such that every pure strategy, if used against an opponent's optimal mixed strategy, yields exactly v to Blue. In such a case we say *all strategies are active* in this game. This notion will be useful for describing the extreme points of the convex set of optimal strategies.

Define J_n to be a row vector which has n elements, each of which is 1; thus

$$J_4 = (1, 1, 1, 1).$$

Now, if X^*, Y^* is a solution of the game having an $n \times n$ payoff matrix A, then all strategies are active if

$$U^* = AY^* = vJ_n' \quad \text{and} \quad L^* = A'X^* = vJ_n'. \tag{3.10}$$

Suppose the payoff matrix is $n \times n$ and nonsingular. If all strategies are active in this game, then the solution is readily obtained. Using (3.10) and solving for X^* and Y^* we have

$$X^* = v(A')^{-1}J_n', \qquad Y^* = vA^{-1}J_n'. \tag{3.11}$$

Since X^* and Y^* are probability distributions, we have

$$J_n X^* = J_n Y^* = 1.$$

Substituting in (3.11) we get

$$(3.12)\qquad \begin{aligned} v &= \frac{1}{J_n(A')^{-1}J'_n} = \frac{1}{J_nA^{-1}J'_n}, \\ X^* &= \frac{(A')^{-1}J'_n}{J_n(A')^{-1}J'_n}, \qquad Y^* = \frac{A^{-1}J'_n}{J_nA^{-1}J'_n}. \end{aligned}$$

Thus, if all strategies are active in a square nonsingular payoff matrix, the solution of the game can be obtained from (3.12). Conversely, if the components of X^* and Y^* in (3.12) are non-negative, then X^*, Y^* is a solution of the game.

Now if A is an $n \times n$ nonsingular matrix, then

$$DA^{-1} = \begin{bmatrix} A_{11} & A_{21} & \dots & A_{n1} \\ A_{12} & A_{22} & \dots & A_{n2} \\ \cdot & \cdot & \dots & \cdot \\ A_{1n} & A_{2n} & \dots & A_{nn} \end{bmatrix} = H,$$

where D is the determinant of A and A_{ji} is the cofactor of the element a_{ij} in A. Recall that $H = DA^{-1}$ is the adjoint of A.

We can express (3.12) in terms of the adjoint matrix as follows:

$$v = \frac{D}{\sum_{i,j} A_{ji}} = \frac{\text{determinant of } A}{\text{sum of all elements of adjoint matrix}},$$

$$x_i^* = \frac{\sum_{j=1}^{n} A_{ji}}{\sum_{i,j} A_{ji}} = \frac{\text{row-sum of elements of adjoint matrix}}{\text{sum of all elements of adjoint matrix}},$$

$$y_j^* = \frac{\sum_{i=1}^{m} A_{ji}}{\sum_{i,j} A_{ji}} = \frac{\text{column-sum of elements of adjoint matrix}}{\text{sum of all elements of adjoint matrix}}.$$

7. OPTIMAL STRATEGIES AS EXTREME POINTS

We have shown that if the payoff matrix is nonsingular and all strategies are active, then the optimal strategies X^*, Y^* and value of game are given by (3.12). Hence X^* is an element of T_1 and Y^* is an element of T_2, where T_1 and T_2 are the convex sets of optimal strategies of Blue and Red, respectively. However, the X^* and Y^* given by (3.12) are special points of T_1 and T_2. We shall show that they are extreme points of the convex sets T_1 and T_2, respectively.

Suppose that X^*, which is given by (3.12), is not an extreme point of T_1. Then we can write

$$X^* = \lambda X_1 + (1 - \lambda)X_2,$$

where X_1 and X_2 are extreme elements of T_1. Since X_1 and X_2 are optimal, it follows that

$$A'X_1 \geq vJ'_n, \tag{3.13}$$

$$A'X_2 \geq vJ'_n. \tag{3.14}$$

It follows that

$$\lambda A'X_1 + (1 - \lambda)A'X_2 = A'[\lambda X_1 + (1 - \lambda)X_2] = A'X^* = vJ'_n.$$

From (3.13) we have that each component of the vector $A'X_1$ is larger than or equal to v. Suppose that some component of X_1 were actually larger than v; then correspondingly $A'X^* > v$ for the same component. This is impossible since we have assumed $A'X^* = vJ'_n$. Therefore $A'X_1 = vJ'_n$. Similarly, $A'X_2 = vJ'_n$.

We have thus shown that

$$A'(X_1 - X_2) = OJ'_n.$$

Since A' is nonsingular, this implies $X_1 = X_2$. Hence

$$X^* = \lambda X_1 + (1 - \lambda)X_2 = X_1,$$

or X^* is an extreme point of T_1.

8. EXTREME POINT WHICH YIELDS SUBMATRIX

Suppose the payoff matrix A is such that there exists a nonsingular $r \times r$ submatrix B for which all strategies are active. Then for the subgame B the game value, v_r, and optimal strategies, X_r^*, Y_r^*, are given by

$$\begin{aligned} v_r &= \frac{1}{J_rB^{-1}J'_r}, \\ X_r^{*\prime} &= \frac{J_rB^{-1}}{J_rB^{-1}J'_r}, \\ Y_r^{*\prime} &= \frac{J_r(B^{-1})'}{J_rB^{-1}J'_r}. \end{aligned} \tag{3.15}$$

Since the problem of solving a game is not affected by interchanging rows or columns, we can suppose without loss of generality that B is situated in the upper left-hand corner of A, i.e.,

$$B = \begin{bmatrix} a_{11} & a_{12} & \cdots & a_{1r} \\ \cdot & \cdot & \cdots & \cdot \\ \cdot & \cdot & \cdots & \cdot \\ \cdot & \cdot & \cdots & \cdot \\ a_{r1} & a_{r2} & \cdots & a_{rr} \end{bmatrix}.$$

Let $X_r^{*\prime} = (x_1^*, x_2^*, \ldots, x_r^*)$ be an optimal strategy of B which is given by (3.15). Now suppose that

$$X^{*\prime} = (x_1^*, x_2^*, \ldots, x_r^*, 0, 0, \ldots, 0)$$

is an optimal strategy in the game with payoff matrix A. We shall show that X^* is an extreme point of T_1. For if we suppose the contrary, then there are distinct strategies $U = (u_1, \ldots, u_m)$ and $W = (w_1, \ldots, w_m)$ of T_1 such that

$$X^* = \lambda U + (1 - \lambda)W, \qquad 0 < \lambda < 1.$$

In terms of the components of the vectors, we have

$$x_i^* = \lambda u_i + (1 - \lambda)w_i, \qquad i = 1, 2, \ldots, m. \tag{3.16}$$

But

$$x_i^* = 0, \qquad i = r + 1, r + 2, \ldots, m.$$

It follows that

$$u_i = w_i = 0, \qquad i = r + 1, r + 2, \ldots, m.$$

Therefore

$$\sum_{i=1}^{r} a_{ij}u_i = \sum_{i=1}^{m} a_{ij}u_i, \qquad j = 1, 2, \ldots, n.$$

Since U is optimal, we have

$$\sum_{i=1}^{m} a_{ij}u_i \geq v, \qquad j = 1, 2, \ldots, n.$$

And hence

$$\sum_{i=1}^{r} a_{ij}u_i \geq v, \qquad j = 1, 2, \ldots, n. \tag{3.17}$$

Similarly, we conclude that

$$\sum_{i=1}^{r} a_{ij}w_i \geq v, \qquad j = 1, 2, \ldots, n. \tag{3.18}$$

From (3.15) we have

$$X_r^{*\prime}B = \frac{J_rB^{-1}B}{J_rB^{-1}J_r'} = vJ_r;$$

hence

$$\sum_{i=1}^{r} a_{ij}x_i^* = v, \qquad j = 1, 2, \ldots, r.$$

Making use of (3.16), we have

$$\lambda \sum_{i=1}^{r} a_{ij}u_i + (1 - \lambda) \sum_{i=1}^{r} a_{ij}w_i = v, \qquad j = 1, 2, \ldots, r. \tag{3.19}$$

From (3.17) and (3.18) and using (3.19), it follows that

$$\sum_{i=1}^{r} a_{ij}u_i = \sum_{i=1}^{r} a_{ij}w_i = v, \qquad j = 1, 2, \ldots, r.$$

Hence

$$U_r'B = W_r'B,$$

which implies

$$(U_r' - W_r')B = 0.$$

Since we have supposed that U and W are distinct vectors, it follows that B is a singular matrix, contrary to hypothesis.

We have therefore shown that an optimal strategy associated with a nonsingular submatrix is an extreme point of T_1.

9. SUBMATRIX WHICH YIELDS EXTREME POINTS

We shall now show that if X^* and Y^* are extreme points of T_1 and T_2, respectively, then there exists a nonsingular submatrix B of A such that

$$\begin{aligned} v &= \frac{1}{J_r B^{-1} J_r'}, \\ X_r^{*\prime} &= \frac{J_r B^{-1}}{J_r B^{-1} J_r'}, \\ Y_r^{*\prime} &= \frac{J_r (B^{-1})'}{J_r B^{-1} J_r'}, \end{aligned} \tag{3.20}$$

where r is the order of B, X_r^* is the vector obtained from X^* by deleting elements corresponding to the rows deleted to obtain B from A, and similarly for Y_r^*.

To prove this we shall take a pair of extreme points, X^*, Y^*, and construct a nonsingular submatrix B satisfying (3.20). For convenience, we shall assume that $v \neq 0$.

First, relabel the rows and columns of A, by using the components of X^* and Y^*, as follows:

$$\begin{aligned} x_i^* &> 0, & 1 &\le i \le m'; \\ x_i^* &= 0, & m' + 1 &\le i \le m; \\ y_j^* &> 0, & 1 &\le j \le n'; \\ y_j^* &= 0, & n' + 1 &\le j \le n. \end{aligned}$$

Now arrange the rows and columns corresponding to $x_i^* = 0$ and $y_j^* = 0$ as follows:

$$\begin{aligned} \sum_{j=1}^{n} a_{ij} y_j^* &= v, & m' + 1 &\le i \le m''; \\ \sum_{j=1}^{n} a_{ij} y_j^* &< v, & m'' + 1 &\le i \le m; \\ \sum_{i=1}^{m} a_{ij} x_i^* &= v, & n' + 1 &\le j \le n''; \\ \sum_{i=1}^{m} a_{ij} x_i^* &> v, & n'' + 1 &\le j \le n. \end{aligned}$$

Define the vectors D_j and C_i as follows:

$$\begin{aligned} D_j &= (a_{1j}, a_{2j}, \ldots, a_{m'j}), & 1 &\le j \le n''; \\ C_i &= (a_{i1}, a_{i2}, \ldots, a_{in'}), & 1 &\le i \le m''. \end{aligned}$$

Now arrange the columns of the matrix for which $n' + 1 \leq j \leq n''$ in such a way that $D_{n'+1}, D_{n'+2}, \ldots, D_t$ are linearly independent of $D_1, D_2, \ldots, D_t$ and $D_{t+1}, \ldots, D_{n''}$ are linearly dependent on $D_1, D_2, \ldots, D_t$. That is, we find some number t such that $n' + 1 \leq t \leq n''$ and such that there *do not exist* numbers $\lambda_1, \lambda_2, \ldots, \lambda_t$ satisfying

$$D_j = \lambda_1 D_1 + \lambda_2 D_2 + \ldots + \lambda_{j-1} D_{j-1} + \lambda_{j+1} D_{j+1} + \ldots + \lambda_t D_t, \qquad n' + 1 \leq j \leq t,$$

and there *do exist* numbers $\lambda_1, \lambda_2, \ldots, \lambda_t$ satisfying

$$D_j = \lambda_1 D_1 + \lambda_2 D_2 + \ldots + \lambda_t D_t, \qquad t + 1 \leq j \leq n''.$$

Similarly, arrange the rows of the matrix for which $m' + 1 \leq i \leq m''$ in such a way that $C_{m'+1}, C_{m'+2}, \ldots, C_s$ are linearly independent of $C_1, C_2, \ldots, C_s$ and $C_{s+1}, C_{s+2}, \ldots, C_{m''}$ are linearly dependent on $C_1, C_2, \ldots, C_s$ where $m' + 1 \leq s \leq m''$. That is, we can find a number s such that if $m' + 1 \leq i \leq s$, there *do not exist* numbers $\mu_1, \mu_2, \ldots, \mu_s$ so that

$$C_i = \mu_1 C_1 + \mu_2 C_2 + \ldots + \mu_{i-1} C_{i-1} + \mu_{i+1} C_{i+1} + \ldots + \mu_s C_s, \tag{3.21}$$

and if $s + 1 \leq i \leq m''$, there do exist $\mu_1, \mu_2, \ldots, \mu_s$ such that

$$C_i = \mu_1 C_1 + \mu_2 C_2 + \ldots + \mu_s C_s. \tag{3.22}$$

We may represent the preceding decomposition of the $m \times n$ matrix as shown in Fig. 5.

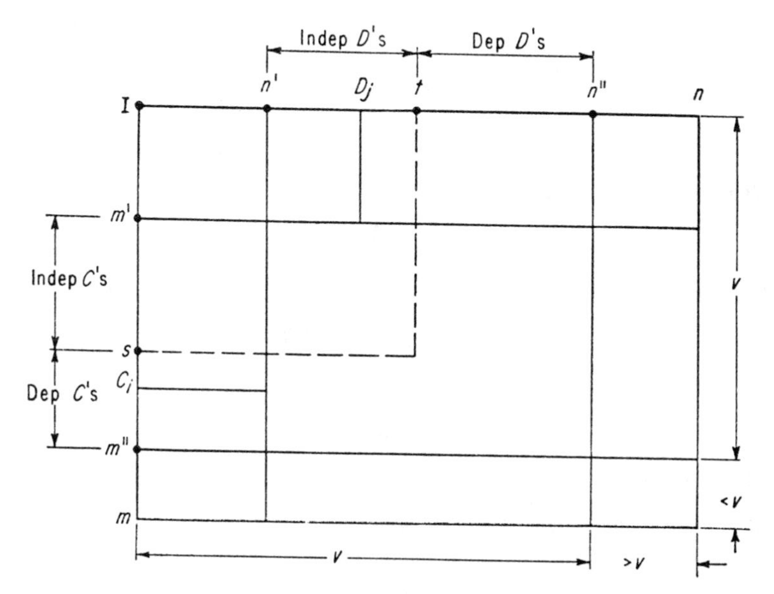

FIGURE 5

We shall now show that the $s \times t$ matrix

$$B = \begin{bmatrix} a_{11} & a_{12} & \dots & a_{1t} \\ . & . & \dots & . \\ a_{s1} & a_{s2} & \dots & a_{st} \end{bmatrix}$$

is nonsingular. Define the vectors

$$B_i = (a_{i1}, a_{i2}, \dots, a_{it}), \qquad 1 \leq i \leq s.$$

Then, if B is singular, we must have dependence among rows or columns. Let us assume the dependence among rows. Then there exist constants c_i, not all zero, such that

$$c_1B_1 + c_2B_2 + \dots + c_sB_s = 0. \tag{3.23}$$

Suppose $c_i \neq 0$ for some i where $m' + 1 \leq i \leq s$. Then we may divide (3.23) by c_i and write

$$B_i = \alpha_1B_1 + \alpha_2B_2 + \dots + \alpha_{i-1}B_{i-1} + \alpha_{i+1}B_{i+1} + \dots + \alpha_sB_s.$$

In particular, we have

$$C_i = \alpha_1C_1 + \alpha_2C_2 + \dots + \alpha_sC_s,$$

which contradicts the assumption of independence of C_i for $m' + 1 \leq i \leq s$. Hence $c_i = 0$ for $m' + 1 \leq i \leq s$.

Thus (3.23) becomes

$$c_1B_1 + c_2B_2 + \dots + c_{m'}B_{m'} = 0,$$

or

$$\sum_{i=1}^{m'} c_ia_{ij} = 0, \qquad 1 \leq j \leq t.$$

Now, since $n' + 1 \leq t \leq n''$ we have

$$\sum_{j=1}^{t} a_{ij}y_j^* = v \qquad \text{for } 1 \leq i \leq m'.$$

Multiplying by c_i, we get

$$c_i \sum_{j=1}^{t} a_{ij}y_j^* = c_iv.$$

Summing over i, we have

$$\sum_{j=1}^{t} \sum_{i=1}^{m'} c_ia_{ij}y_j^* = v \sum_{i=1}^{m'} c_i,$$

or

$$0 = v \sum_{i=1}^{m'} c_i.$$

Since $v \neq 0$, it follows that

$$\sum_{i=1}^{m'} c_i = 0. \tag{3.24}$$

Now define the following m component vectors,

$$X_\epsilon = (x_1^* + \epsilon c_1, x_2^* + \epsilon c_2, \ldots, x_{m'}^* + \epsilon c_{m'}, 0, \ldots, 0),$$

$$X_{-\epsilon} = (x_1^* - \epsilon c_1, x_2^* - \epsilon c_2, \ldots, x_{m'}^* - \epsilon c_{m'}, 0, \ldots, 0),$$

where ϵ is arbitrary and c_i satisfies (3.24). Then we can find an ϵ such that

$$x_i^* + \epsilon c_i \geq 0, \qquad 1 \leq i \leq m';$$

$$x_i^* - \epsilon c_i \geq 0, \qquad 1 \leq i \leq m'.$$

Hence

$$\sum_{i=1}^{m} (x_i^* + \epsilon c_i) = \sum_{i=1}^{m} x_i^* + \epsilon \sum_{i=1}^{m} c_i = 1 + \epsilon \sum_{i=1}^{m'} c_i = 1,$$

$$\sum_{i=1}^{m} (x_i^* - \epsilon c_i) = 1 - \epsilon \sum_{i=1}^{m'} c_i = 1.$$

Therefore X_ϵ and $X_{-\epsilon}$ are mixed strategies for Blue.

Now for all j, we have

$$\sum_{i=1}^{m} a_{ij}(x_i^* \pm \epsilon c_i) = \sum_{i=1}^{m} a_{ij}x_i^* \geq v.$$

Therefore X_ϵ and $X_{-\epsilon}$ are optimal strategies for Blue. But

$$X^* = \tfrac{1}{2}(X_\epsilon + X_{-\epsilon}),$$

which contradicts the assumption that X^* is an extreme point of T_1. Hence B is nonsingular. This implies that B is a square matrix, or $s = t$. Further, it is clear from our construction of s and t that for the B matrix all strategies are active. Therefore, (3.20) will hold for the B matrix.

10. DETERMINING THE SETS OF OPTIMAL STRATEGIES

The sets of optimal strategies, T_1 for Blue and T_2 for Red, are convex sets. Hence they are determined by the extreme points of the sets. Now for each pair of extreme points, one from T_1 and one from T_2, there is a nonsingular submatrix whose solution has all strategies active. Further, every nonsingular submatrix, whose solution has all strategies active and which is also a solution of the original matrix, determines a pair of extreme points of T_1 and T_2. Since there are only a finite number of submatrices in each matrix, it follows that there are only a finite number of extreme points of T_1 and T_2. Therefore the sets T_1 and T_2 are polyhedral sets spanned by a finite number of vertices.

We can find all the extreme points of T_1 and T_2 by examining all the submatrices of the payoff matrix.

Example. Let us determine the sets of optimal strategies T_1 and T_2 for the following game:

$$A = \begin{bmatrix} 3 & 5 & 3 \\ 4 & -3 & 2 \\ 3 & 2 & 3 \end{bmatrix}.$$

First, we test for strict dominance to see if there are any poor strategies in this game. We find no strict dominance, although there are several instances of nonstrict dominance.

To obtain all the extreme points of T_1 and T_2 we examine all square submatrices that can be formed. There are nine 1×1 submatrices, nine 2×2 submatrices, and one 3×3 matrix. Examining 1×1 submatrices is equivalent to testing A for saddle-points. We find a saddle-point at $i = 1$, $j = 3$. Therefore $(1, 0, 0)$ is an extreme point of T_1 and $(0, 0, 1)$ is an extreme point of T_2. The corresponding nonsingular submatrix is (3).

Of the nine 2×2 submatrices, the following three,

$$\begin{bmatrix} -3 & 2 \\ 2 & 3 \end{bmatrix}, \quad \begin{bmatrix} 5 & 3 \\ -3 & 2 \end{bmatrix}, \quad \begin{bmatrix} 4 & -3 \\ 3 & 2 \end{bmatrix},$$

exhibit strict dominance and therefore cannot have all strategies active. The matrix

$$\begin{bmatrix} 3 & 3 \\ 3 & 3 \end{bmatrix}$$

is singular and hence need not be considered. For each of the remaining five matrices,

$$\begin{bmatrix} 5 & 3 \\ 2 & 3 \end{bmatrix}, \quad \begin{bmatrix} 4 & 2 \\ 3 & 3 \end{bmatrix}, \quad \begin{bmatrix} 3 & 3 \\ 4 & 2 \end{bmatrix}, \quad \begin{bmatrix} 3 & 5 \\ 3 & 2 \end{bmatrix}, \quad \begin{bmatrix} 3 & 5 \\ 4 & -3 \end{bmatrix},$$

all strategies are active. Their solutions can be computed by inverting the corresponding matrices. For example, to solve the second matrix of the five, we have the following computations:

$$B = \begin{bmatrix} 4 & 2 \\ 3 & 3 \end{bmatrix}, \quad B^{-1} = \frac{1}{6}\begin{bmatrix} 3 & -2 \\ -3 & 4 \end{bmatrix}, \quad v = \frac{1}{J_2 B^{-1} J_2'} = \frac{6}{2} = 3,$$

$$Y^* = vB^{-1}J_2' = \frac{1}{2}\begin{bmatrix} 3 & -2 \\ -3 & 4 \end{bmatrix}\begin{bmatrix} 1 \\ 1 \end{bmatrix} = \begin{bmatrix} \frac{1}{2} \\ \frac{1}{2} \end{bmatrix},$$

$$X^* = v(B')^{-1}J_2 = \frac{1}{2}\begin{bmatrix} 3 & -3 \\ -2 & 4 \end{bmatrix}\begin{bmatrix} 1 \\ 1 \end{bmatrix} = \begin{bmatrix} 0 \\ 1 \end{bmatrix}.$$

If we test $(0, 0, 1)$, $(\frac{1}{2}, 0, \frac{1}{2})$ for a solution of the original game A, we find

$$\max_i \sum_{j=1}^{3} a_{ij}y_j^* = 3, \qquad \min_j \sum_{i=1}^{3} a_{ij}x_i^* = 2.$$

Therefore $(0, 0, 1)$, $(\frac{1}{2}, 0, \frac{1}{2})$ is not a solution of A. Of the five 2×2 matrices we find that the two matrices,

$$\begin{bmatrix} 5 & 3 \\ 2 & 3 \end{bmatrix} \qquad \begin{bmatrix} 3 & 3 \\ 4 & 2 \end{bmatrix},$$

yield solutions which are also solutions of the game. They are, respectively,

$$(\tfrac{1}{3}, 0, \tfrac{2}{3}), \quad (0, 0, 1);$$

$$(1, 0, 0), \quad (\tfrac{1}{2}, 0, \tfrac{1}{2}).$$

Finally, the 3×3 full matrix must be tested for a solution. We have:

$$A = \begin{bmatrix} 3 & 5 & 3 \\ 4 & -3 & 2 \\ 3 & 2 & 3 \end{bmatrix}, \qquad A^{-1} = -\frac{1}{18}\begin{bmatrix} -13 & -9 & 19 \\ -6 & 0 & 6 \\ 17 & 9 & -29 \end{bmatrix},$$

$$v = \frac{1}{J_3A^{-1}J_3'} = \frac{-18}{-6} = 3,$$

$$vA^{-1}J_3' = \begin{bmatrix} \frac{1}{2} \\ 0 \\ \frac{1}{2} \end{bmatrix} = Y^*, \qquad v(A')^{-1}J_3' = \begin{bmatrix} \frac{1}{3} \\ 0 \\ \frac{2}{3} \end{bmatrix} = X^*.$$

Thus the 3×3 matrix yields an extreme point. But it is identical to one of the extreme points previously determined in a 2×2 submatrix. Therefore T_1 is the line joining $(1, 0, 0)$ and $(\frac{1}{3}, 0, \frac{2}{3})$. T_2 is the line joining $(0, 0, 1)$ and $(\frac{1}{2}, 0, \frac{1}{2})$. Every optimal strategy of Blue can be expressed by

$$\alpha(1, 0, 0) + (1 - \alpha)(\tfrac{1}{3}, 0, \tfrac{2}{3}), \qquad 0 \leq \alpha \leq 1.$$

Every optimal strategy of Red can be expressed by

$$\beta(0, 0, 1) + (1 - \beta)(\tfrac{1}{2}, 0, \tfrac{1}{2}), \qquad 0 \leq \beta \leq 1.$$

11. GEOMETRY OF SOLUTIONS

We have described the properties of each optimal strategy. We also have described how to construct the set of optimal strategies, which is a convex set. In this section we shall describe additional geometric relationships that exist among the sets of optimal strategies. These relationships can be used to check the completeness of the sets of solutions.

We shall need to introduce some additional terms and symbols associated with the set of solutions. To clarify these terms and symbols we shall apply them to the particular game whose payoff matrix is defined by

$$A = \begin{matrix} & R_1 & R_2 & R_3 \\ B_1 & -1 & 3 & -3 \\ B_2 & 2 & 0 & 3 \\ B_3 & 2 & 1 & 0 \end{matrix}.$$

Let the set of pure strategies for Blue be denoted by $I = \{1, 2, \ldots, m\}$ and for Red by $J = \{1, 2, \ldots, n\}$. In our example $I = \{B_1, B_2, B_3\}$, $J = \{R_1, R_2, R_3\}$.

The set of mixed strategies for Blue can be represented by points $X = (x_1, x_2, \ldots, x_m)$ in m-dimensional space whose coordinates are non-negative and whose sum is 1. The set of such points defines a simplex $S(I)$. Red's set of mixed strategies is a simplex $S(J)$. For the 3×3 matrix of our example, $S(I)$ is an equilateral triangle with sides of length $\sqrt{2}$ (Fig. 6).

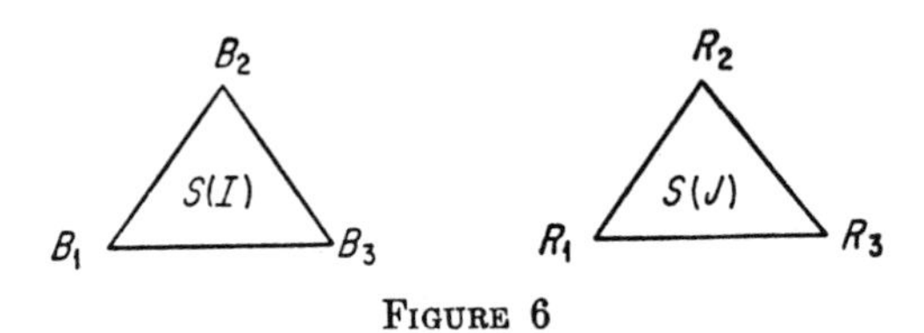

FIGURE 6

The vertices of the triangle represent the pure strategies. Similarly, $S(J)$ is an equilateral triangle. Every point in the triangle represents a mixed strategy.

We let $X(A)$ be the set of optimal mixed strategies of Blue—i.e., for each $X \varepsilon X(A)$ we have

$$\min_Y X'AY = v.$$

Let $Y(A)$ be the set of optimal mixed strategies of Red—i.e., for $Y \varepsilon Y(A)$ we have

$$\max_X X'AY = v.$$

In our example, it can be verified that Blue has a unique optimal strategy—$(\frac{1}{3}, \frac{2}{3}, 0)$. However, Red's set of optimal strategies has two extreme points—$(0, \frac{2}{3}, \frac{1}{3})$ and $(\frac{1}{5}, \frac{3}{5}, \frac{1}{5})$. Thus $X(A)$ is the point whose coordinates are $(\frac{1}{3}, \frac{2}{3}, 0)$ and $Y(A)$ is the line segment, whose end points are $(0, \frac{2}{3}, \frac{1}{3})$ and $(\frac{1}{5}, \frac{3}{5}, \frac{1}{5})$.

With each mixed strategy X of Blue we can associate a set $I_1(X)$ which is the set of pure strategies that have positive weights in X. Thus

$$I_1(0, 1, 0) = \{B_2\} \quad \text{and} \quad I_1(\tfrac{1}{3}, \tfrac{2}{3}, 0) = \{B_1, B_2\}.$$

Let $I_2(Y)$ be the set of pure strategies of Blue which yield v when played against Y. In our example, it can be verified that $I_2(0, \frac{2}{3}, \frac{1}{3}) = \{B_1, B_2\}$. In a similar manner we may define for Red the sets $J_1(Y)$ and $J_2(X)$. Thus

$$J_1(\tfrac{1}{5}, \tfrac{3}{5}, \tfrac{1}{5}) = \{R_1, R_2, R_3\}$$

and

$$J_2(\tfrac{1}{3}, \tfrac{2}{3}, 0) = \{R_1, R_2, R_3\}\cdot$$

Effective strategies. It is evident that for all X interior to $X(A)$ the set $I_1(X)$ is the same, so let us denote this set by I_1. Also, for all Y interior to $Y(A)$ the set $I_2(Y)$ is the same; we shall denote it by I_2. Further, it can be shown that the two sets are identical, or

$$I_1 = I_2.$$

The identical sets I_1 and I_2 are called the *effective strategies* for Blue. They have two important properties: First, they are the strategies which appear with positive weights in every interior solution; secondly, they yield the value of the game against every interior optimal mixed strategy of Red. With a similar definition for Red's strategies, it follows that

$$J_1 = J_2,$$

and they are the effective strategies for Red.

In our example, Blue has a unique optimal strategy $X = (\frac{1}{3}, \frac{2}{3}, 0)$. Thus $I_1(X) = \{B_1, B_2\}$. Now the set $Y(A)$ consists of strategies of the form

$$Y = \left[\frac{1-\alpha}{5}, \frac{2}{3}\alpha + \frac{3}{5}(1-\alpha), \frac{\alpha}{3} + \frac{1}{5}(1-\alpha)\right],$$

where $0 \leq \alpha \leq 1$. Now for each Y interior to $Y(A)$ we obtain the value of the game, 1, for strategies B_1 and B_2. Thus $I_2(Y) = \{B_1, B_2\} = I_2 = I_1$. Similarly $J_1(Y) = \{R_1, R_2, R_3\} = J_1$ and $J_2(X) = \{R_1, R_2, R_3\} = J_2 = J_1$. Two strategies, B_1 and B_2, are effective for Blue and all three strategies are effective for Red.

Essential subgame. If we construct a payoff matrix A_1 having the set I_1 for its rows and J_1 for its columns, we obtain the *essential submatrix* of A. The value of the essential subgame is the same as the value of the game A. Further, $X(A)$ will be a subset of $X(A_1)$ and $Y(A)$ will be a subset of $Y(A_1)$.

The essential submatrix in our example is determined by $I_1 = \{B_1, B_2\}$ and $J_1 = \{R_1, R_2, R_3\}$, or

$$A_1 = \begin{array}{c} \\ B_1 \\ B_2 \end{array} \begin{array}{c} \begin{array}{ccc} R_1 & R_2 & R_3 \end{array} \\ \begin{bmatrix} -1 & 3 & -3 \\ 2 & 0 & 3 \end{bmatrix} \end{array}.$$

Solving A_1 we get $X(A_1) = (\frac{1}{3}, \frac{2}{3})$ and $Y(A_1)$ is the line connecting $(0, \frac{2}{3}, \frac{1}{3})$ and $(\frac{1}{2}, \frac{1}{2}, 0)$. The value of the game A_1 is 1. Now $Y(A)$, which is the line with end points $(0, \frac{2}{3}, \frac{1}{3})$ and $(\frac{1}{5}, \frac{3}{5}, \frac{1}{5})$, is clearly a subset of $Y(A_1)$.

Dimension of solutions. The sets $X(A)$ and $Y(A)$ are convex polyhedra. Let $S(I_1)$ be the simplex of the effective strategies of Blue. Then $S(I_1)$ is the smallest closed simplicial face of $S(I)$ that contains $X(A)$. Similarly, $S(J_1)$ is the smallest closed simplicial face of $S(J)$ that contains $Y(A)$.

Since $I_1 = \{B_1, B_2\}$ in our example, $S(I_1)$ is the line joining B_1 and B_2 of $S(I)$. Since $J_1 = \{R_1, R_2, R_3\}$, then $S(J_1) = S(J)$. The simplex $S(I_1)$

has dimension 1 and $S(J_1)$ has dimension 2. Since $S(A)$ consists of one point, it is therefore of dimension 0. The dimension of $Y(A)$ is 1. In general, the dimensions of these various sets are related as follows:

$$\begin{aligned}\dim S(I_1) - \dim X(A) &= \dim S(J_1) - \dim Y(A)\\ &= \begin{cases}\text{rank } A_1 & \text{if } v = 0\\ \text{rank } A_1 - 1 & \text{if } v \neq 0.\end{cases}\end{aligned}$$

In our example, we have rank $A_1 = 2$, $v = 1$, and dim $Y(A) = 1$. Substituting in the above we have

$$1 - 0 = 2 - 1 = 2 - 1.$$

12. TARGET SELECTION—FOR ATTACK AND DEFENSE

The target selection problem frequently appears in military situations. In its most general form the problem may be described as follows: Suppose the Attack and Defense each have a fixed quantity of resources to allocate among a series of targets of different values. How should the Attack and Defense select the targets for the allocations?

Let us examine the simplest of the target selection problems. The Attack has one unit and the Defense has one unit; each unit is to be allocated to some target. Which target should receive the allocation?

The game may be described as follows:

Targets. There are n targets which we label $T_1, T_2, \ldots, T_n$. We assume that these targets have values $a_1, a_2, \ldots, a_n$, respectively, which are ordered as follows:

$$a_1 > a_2 > \ldots > a_n > 0.$$

Attack. The Attack, Blue, has one attacking unit to allocate to some one of the n targets.

Defense. The Defense, Red, has one unit of defense to allocate to some one of his targets. It is assumed that the unit of defense has a defense potential p—i.e., if an attack is made on a defended target, then with probability p, the Attack fails to destroy the target. Hence $1 - p$ is the probability that Blue is successful in destroying this attacked target.

Strategy. A strategy for Blue is a choice of a target for attack. A strategy for Red is a choice of a target for defense. Hence each player has n strategies.

Payoff. We shall assume that if an attack is made on an undefended target, T_k, then the payoff to Blue is the value, a_k, of that target. However, if an attack is made on a defended target, T_i, then the payoff is $(1 - p)a_i$, the expected damage to the target. Therefore, the payoff matrix is the following:

TARGET SELECTION PAYOFF

Target defended

		T_1	T_2	$\dots$	T_n
	T_1	$(1-p)a_1$	a_1	$\dots$	a_1
	T_2	a_2	$(1-p)a_2$	$\dots$	a_2
Target	.	.	.	$\dots$	.
attacked	.	.	.	$\dots$	.
	.	.	.	$\dots$	.
	T_n	a_n	a_n	$\dots$	$(1-p)a_n$

Because of the special form of the payoff matrix, we shall solve this game by guessing a solution and then verifying it. However, to make a good guess it is necessary to make some preliminary investigations.

It is reasonable to expect the solution to have the property that each player randomizes over the same high-valued targets. That is, let us assume that X^*, Y^* is of the following form:

$$X^* = (x_1, x_2, \dots, x_t, 0, 0, 0),$$
$$Y^* = (y_1, y_2, \dots, y_t, 0, 0, 0),$$

where

$$x_i > 0, \qquad i \le t;$$
$$x_i = 0, \qquad i > t;$$
$$y_j > 0, \qquad j \le t;$$
$$y_j = 0, \qquad j > t;$$

where the value of t is to be determined.

Since $y_1 > 0$, $y_2 > 0$, it follows that

$$(1-p)a_1x_1 + a_2x_2 + \dots + a_tx_t = v,$$
$$a_1x_1 + (1-p)a_2x_2 + \dots + a_tx_t = v.$$

Subtracting the first equation from the second, we get

$$pa_1x_1 - pa_2x_2 = 0,$$

or

$$x_2 = \frac{a_1x_1}{a_2}.$$

In general, we have

$$x_i = \frac{a_1x_1}{a_i} \qquad \text{for } i \le t.$$

However,

$$\sum_{i=1}^{n} x_i = \sum_{i=1}^{t} x_i = \sum_{i=1}^{t} \frac{a_1x_1}{a_i} = 1.$$

Let us define

$$A_k = \sum_{i=1}^{k} \frac{1}{a_i}.$$

We get

$$x_1 = \frac{1}{a_1 A_t}.$$

Therefore X^* must be such that

$$x_i = \frac{1}{a_i A_t} \qquad \text{for } i \leq t,$$
$$= 0 \qquad \text{for } i > t.$$

To obtain the form of Y^* we note that $x_1 > 0$ and $x_2 > 0$. Therefore

$$(1 - p)a_1 y_1 + a_1 y_2 + \ldots + a_1 y_t = v,$$
$$a_2 y_1 + (1 - p)a_2 y_2 + \ldots + a_2 y_t = v.$$

From this pair of equations and since Y^* is a mixed strategy, we get

$$y_2 = \frac{1}{p}\left[1 - \frac{a_1}{a_2}(1 - py_1)\right].$$

Using the fact that $x_1 > 0$ and $x_i > 0$ for $i \leq t$ we obtain, by a similar argument,

$$y_j = \frac{1}{p}\left[1 - \frac{a_1}{a_j}(1 - py_1)\right] \qquad \text{for } j \leq t.$$

But

$$\sum_{j=1}^{t} y_j = 1.$$

This yields Y^* of the form

$$y_j = \frac{1}{p}\left[1 - \frac{t-p}{a_j A_t}\right] \qquad \text{for } j \leq t,$$
$$= 0 \qquad \text{for } j > t.$$

We also obtain a guess of the value of the game, since we must have

$$v = (1 - p)a_1 x_1 + a_2 x_2 + \ldots + a_t x_t$$
$$= \sum_{i=1}^{t} a_i x_i - pa_1 x_1$$
$$= \frac{t - p}{A_t}.$$

Verification. We shall now prove, by verifying, that

(i) The optimal strategy for Attack is

$$x_i = \frac{1}{a_i A_t} \qquad \text{for } i \leq t,$$
$$= 0 \qquad \text{for } i > t.$$

(ii) The optimal strategy for Defense is

$$y_j = \frac{1}{p}\left(1 - \frac{1}{a_j}\frac{t-p}{A_t}\right) \qquad \text{for } j \leq t,$$
$$= 0 \qquad \text{for } j > t.$$

(iii) The value of the game is

$$v = \frac{t-p}{A_t} = \max_{k \leq n} \frac{k-p}{A_k}.$$

Note that the value of t is defined by the location of the maximum of $(k-p)/A_k$.

Let us verify that the strategy above is optimal for the Attack. Clearly $x_i \geq 0$ and $\sum_{i=1}^{n} x_i = 1$. Hence (i) defines mixed strategy. Computing expectations, we have

$$\sum_{i=1}^{n} a_{ij}x_i = \sum_{\substack{i=1\\ i\neq j}}^{n} a_i x_i + a_j(1-p)x_j$$

$$= \begin{cases} \sum_{j\neq i \leq t} a_i \dfrac{1}{a_i A_t} + a_j(1-p)\dfrac{1}{a_j A_t} & \text{for } j \leq t, \\ \sum_{j \neq i \leq t} a_i \dfrac{1}{a_i A_t} & \text{for } j > t, \end{cases}$$

$$= \begin{cases} \dfrac{t-p}{A_t} & \text{for } j \leq t, \\ \dfrac{t}{A_t} & \text{for } j > t. \end{cases}$$

Therefore,

$$\sum_{i=1}^{n} a_{ij}x_i \geq \frac{t-p}{A_t} \qquad \text{for all } j.$$

Now let us verify that (ii) is optimal for Defense. Clearly $\sum_{j=1}^{n} y_j = 1$. That $y_j \geq 0$ will be verified later. Now compute expectations

$$\sum_{j=1}^{n} a_{ij}y_j = \sum_{i\neq j\leq n} a_i y_j + (1-p)a_i y_i$$
$$= a_i \sum_{i \neq j \leq n} y_j + (1-p)a_i y_i$$
$$= a_i(1 - py_i)$$
$$= \begin{cases} \dfrac{t-p}{A_t} & \text{if } i \leq t, \\ a_i & \text{if } i > t. \end{cases}$$

This implies that if Y^*, as defined above, is a solution, we need

$$a_i \leq \frac{t - p}{A_t} \qquad \text{for } i > t.$$

In particular,

$$a_{t+1} \leq \frac{t - p}{A_t},$$

or

$$p \leq t - a_{t+1}A_t = t + 1 - a_{t+1}A_{t+1}.$$

Also, since we require $y_j \geq 0$, we get $p \geq t - a_tA_t$.

The preceding leads us to study the function

$$\phi_k = k - a_kA_k.$$

Let us compute

$$\begin{aligned} \phi_{k+1} - \phi_k &= k + 1 - a_{k+1}A_{k+1} - k + a_kA_k \\ &= 1 - a_{k+1}A_{k+1} + a_kA_k \\ &> 1 - a_{k+1}\left(A_k + \frac{1}{a_{k+1}}\right) + a_{k+1}A_k = 0. \end{aligned}$$

Thus ϕ is a strictly increasing function of k. Since $\phi_1 = 0$, and if we define $\phi_{n+1} = \infty$, then it follows that there exists exactly one positive integer $t \leq n$ such that

$$\phi_t \leq p < \phi_{t+1}.$$

For this integer t, we have

$$t - a_tA_t \leq p < t + 1 - a_{t+1}A_{t+1}.$$

This gives us the inequality

$$a_{t+1} < \frac{t - p}{A_t} \leq a_t.$$

This inequality combined with

$$a_1 > a_2 > \ldots > a_n$$

yields

$$\frac{t - p}{A_t} \leq a_i \qquad \text{for } i \leq t,$$

$$\frac{t - p}{A_t} > a_i \qquad \text{for } i > t.$$

To complete the verification we need to show that $y_j \geq 0$ for $j \leq t$. This is exactly the statement

$$\frac{t - p}{A_t} \leq a_j \qquad \text{for } j \leq t$$

which has been shown above.

Recall that the value of t was defined by the inequality

$$\phi_t \leq p \leq \phi_{t+1}.$$

We shall now show that t is also given by

$$\max_{k \le n} \frac{k - p}{A_k} = \frac{t - p}{A_t}.$$

The function

$$f(k) = \frac{k - p}{A_k}$$

has a maximum, say at $k = u$. Hence we have

$$f(u + 1) \le f(u),$$
$$f(u - 1) \le f(u).$$

Evaluating the function at u, $u - 1$, and $u + 1$, these inequalities yield

$$u - a_u A_u \le p \le u + 1 - a_{u+1} A_{u+1}.$$

This is exactly the inequality satisfied by the integer t. Hence $u = t$.

13. SOLUTION OF THE GAME "LE HER"

This game was described in Example 3, Chapter 1. If we define the payoff to be 1 if the dealer wins, and 0 if the receiver wins, then we can compute the payoff matrix associated with the $2^{13} \times 2^{13}$ ways of playing "le Her." For example, if the strategy selected by the receiver is to change any card which is 6 or under and to hold any card which is 7 or over, and if the strategy selected by the dealer is to change any card which is 8 or under and to hold any card which is 9 or over, then the payoff corresponding to this way of playing the game is obtained from the matrix

		Receiver												
		C 1	C 2	C 3	C 4	C 5	C 6	S 7	S 8	S 9	S 10	S 11	S 12	S 13
	C 1	50	50	50	50	50	50	23	19	15	11	7	3	0
	C 2	43	50	50	50	50	50	23	19	15	11	7	3	0
	C 3	39	39	50	50	50	50	23	19	15	11	7	3	0
	C 4	35	35	35	50	50	50	23	19	15	11	7	3	0
	C 5	31	31	31	31	50	50	23	19	15	11	7	3	0
	C 6	27	27	27	27	27	50	23	19	15	11	7	3	0
Dealer	C 7	23	23	23	23	23	23	26	19	15	11	7	3	0
	C 8	19	19	19	19	19	19	26	22	15	11	7	3	0
	S 9	15	15	15	15	15	15	50	50	50	0	0	0	0
	S 10	11	11	11	11	11	11	50	50	50	50	0	0	0
	S 11	7	7	7	7	7	7	50	50	50	50	50	0	0
	S 12	3	3	3	3	3	3	50	50	50	50	50	50	0
	S 13	50	50	50	50	50	50	50	50	50	50	50	50	50

where the elements in the matrix represent 50 times the probability of the dealer's winning for each of the possible ways that the cards may be dealt.

Now the probability that the particular pairs of cards will be dealt is given by

$$\tfrac{4}{52} \cdot \tfrac{4}{51} = \tfrac{4}{663} \quad \text{for nondiagonal elements,}$$

$$\tfrac{4}{52} \cdot \tfrac{3}{51} = \tfrac{3}{663} \quad \text{for diagonal elements.}$$

Therefore we need to multiply diagonal elements of the matrix above by 3/(663)(50) and nondiagonal elements by 4/(663)(50) and add the 169 results, which gives 16,146/(663)(50), the payoff for this pair of strategies.

It is not necessary to compute the payoff for the entire $2^{13} \times 2^{13}$ matrix, for it is apparent that any strategy in which exactly r changes of cards appear is dominated by a strategy in which the first r cards are changes and the remaining $13 - r$ are stays. This follows from a dominance argument. It is also intuitively evident, because if, for example, it is advisable to stay with a 3 and to change a 4, it is certainly more advisable to change a 3. This reduces the matrix to 14 × 14. Finally, the 14 × 14 matrix can be reduced to a 2 × 2 matrix, again by a dominance argument. The resulting 2 × 2 matrix is the following:

		Receiver Hold 7 and over	*Receiver* Change 7 and under
Dealer	Hold 8 and over	$\frac{16{,}182}{33{,}150}$	$\frac{16{,}122}{33{,}150}$
Dealer	Change 8 and under	$\frac{16{,}146}{33{,}150}$	$\frac{16{,}182}{33{,}150}$

Solving the 2 × 2 game, we obtain the optimal strategies

for the Dealer: $\begin{bmatrix} \frac{3}{8} \\ \frac{5}{8} \end{bmatrix}$,

for the Receiver: $\begin{bmatrix} \frac{5}{8} \\ \frac{3}{8} \end{bmatrix}$,

and the value of the game is

for the Dealer: 0.487,

for the Receiver: 0.513.

14. SOLUTION OF THE GAME OF "MORRA"

This game was described in Example 2, Chapter 1. Since the game is symmetric, the two players will have the same strategies. The solution can be obtained by examining square submatrices and using dominance arguments. There are four extreme points and they are close to one another:

$$(0, 0, \tfrac{5}{12}, 0, \tfrac{4}{12}, 0, \tfrac{3}{12}, 0, 0),$$
$$(0, 0, \tfrac{16}{37}, 0, \tfrac{12}{37}, 0, \tfrac{9}{37}, 0, 0),$$
$$(0, 0, \tfrac{20}{47}, 0, \tfrac{15}{47}, 0, \tfrac{12}{47}, 0, 0),$$
$$(0, 0, \tfrac{25}{61}, 0, \tfrac{20}{61}, 0, \tfrac{16}{61}, 0, 0).$$

It is also interesting to note that the essential submatrix has rank zero; it is the matrix

$$\begin{bmatrix} 0 & 0 & 0 \\ 0 & 0 & 0 \\ 0 & 0 & 0 \end{bmatrix}.$$

15. RECONNAISSANCE AS A GAME OF STRATEGY

The advisability of reconnaissance before attack can be investigated by considering the problem as a game of strategy. The purpose of reconnaissance is to obtain information about the enemy's strategic intentions. Having the information, the attacker can more effectively plan his attack. However, the attempt to obtain information is costly and it may not succeed if the enemy invokes effective countermeasures. This conflict of interest, whether or not to reconnoiter, can be resolved within the framework of a game of strategy.

To illustrate the essential principles, we shall solve a hypothetical reconnaissance problem associated with a small number of strategies for each side. No additional principles are introduced if the players have a large number of strategies. Thus we shall assume that the attacker and defender have two strategies each.

Let us assume that the attacker, Blue, wishes to seize a defended enemy position. For simplicity, let us assume that he has two courses of action, his two strategies:

① attack with his entire force;
② attack with part of his force, leaving the remainder as reserves and a rear guard in case the enemy "outflanks" him.

The defender, Red, is assumed to have two possible courses of action, his two strategies:

[1] defend with his entire force the objective of the attacker;
[2] defend with part of his force, and send the remainder to "outflank" the enemy and attack the enemy from the rear.

There are four possible outcomes of the above courses of action. They can be summarized by the following 2×2 matrix:

$$A = \begin{matrix} & \boxed{1} & \boxed{2} \\ ① & a_{11} & a_{12} \\ ② & a_{21} & a_{22} \end{matrix},$$

where, for example, a_{21} represents the value to Blue if he attacks with part of his force and Red defends with his entire force.

Suppose that the outcomes are such that if Red uses strategy $\boxed{1}$, then Blue would prefer to use strategy ①, i.e., $a_{11} > a_{21}$, and if Red uses strategy $\boxed{2}$, then Blue would prefer to use strategy ②. Clearly, the attacker could benefit from a knowledge of the defender's intentions. Thus, the attacker might find it profitable to send out a detachment of men to reconnoiter in an attempt to discover the plans of the defender. In order to defend himself against such possible action the defender may take countermeasures.

Now if the attacker decides to reconnoiter he must sacrifice some of his attacking forces. If the defender decides to take countermeasures he must sacrifice some of his defensive forces. Let c be the cost to the attacker in in order to reconnoiter and d be the cost to the defender in order to use countermeasures.

The decisions of the attacker, whether or not to reconnoiter, and of the defender, whether or not to use countermeasures, have increased the number of strategies available to the two sides. Whereas the original game had only two strategies for each side, the resulting reconnaissance game will be seen to have 16 strategies for the attacker and 4 for the defender. Of course, many of the strategies will turn out to be poor strategies. But in establishing the game we must enumerate all of them.

A strategy for the attacker will be a set of instructions which tell him how to act taking into account the information he may receive. A convenient way to represent symbolically a strategy for the attacker is by an ordered sequence of numbers $(u; x, y, z)$ in which each letter takes on the value 1 or 2, and they have the following meanings:

u = 1. Reconnoiter.
u = 2. Do not reconnoiter.
x = 1. Play strategy ① if no information is received about the defender.
x = 2. Play strategy ② if no information is received about the defender.
y = 1. Play strategy ① if the information indicates that the defender is using strategy $\boxed{1}$.
y = 2. Play strategy ② if the information indicates that the defender is using strategy $\boxed{1}$.
z = 1. Play strategy ① if the information indicates that the defender is using strategy $\boxed{2}$.
z = 2. Play strategy ② if the information indicates that the defender is using strategy $\boxed{2}$.

For example, the strategy $(1; 1, 2, 1)$ instructs the attacker to reconnoiter, and, if no information is obtained, to play strategy ①—attack with all his forces; if the information indicates that the enemy is using strategy $\boxed{1}$ play strategy ②; if the information indicates that the enemy is using strategy $\boxed{2}$ play strategy ①. There will be $2^4 = 16$ different strategies for the attacker. Some of them are redundant, and they will show up as such in the payoff matrix.

A strategy for the defender is an ordered sequence $(s; t)$ where each letter takes on the value 1 or 2 with the following meanings:

$s = 1$. Take countermeasures.
$s = 2$. Take no countermeasures.
$t = 1$. Use strategy $\boxed{1}$.
$t = 2$. Use strategy $\boxed{2}$.

Thus $(1; 2)$ means that Red takes countermeasures and uses strategy $\boxed{2}$, i.e., defends with a partial force.

With this notation for strategies, the 64 possible outcomes of the game can be summarized by the payoff matrix A_1 given by:

RECONNAISSANCE PAYOFF

$$A_1 = \begin{array}{c} \\ (1;1,1,1) \\ (1;1,1,2) \\ (1;1,2,1) \\ (1;1,2,2) \\ (1;2,1,1) \\ (1;2,1,2) \\ (1;2,2,1) \\ (1;2,2,2) \\ (2;1,1,1) \\ (2;1,1,2) \\ (2;1,2,1) \\ (2;1,2,2) \\ (2;2,1,1) \\ (2;2,1,2) \\ (2;2,2,1) \\ (2;2,2,2) \end{array} \begin{array}{c} \begin{array}{cccc} (1;1) & (1;2) & (2;1) & (2;2) \end{array} \\ \left[\begin{array}{llll} a_{11} - c + d & a_{12} - c + d & a_{11} - c & a_{12} - c \\ a_{11} - c + d & a_{12} - c + d & a_{11} - c & a_{22} - c \\ a_{11} - c + d & a_{12} - c + d & a_{21} - c & a_{12} - c \\ a_{11} - c + d & a_{12} - c + d & a_{21} - c & a_{22} - c \\ a_{21} - c + d & a_{22} - c + d & a_{11} - c & a_{12} - c \\ a_{21} - c + d & a_{22} - c + d & a_{11} - c & a_{22} - c \\ a_{21} - c + d & a_{22} - c + d & a_{21} - c & a_{12} - c \\ a_{21} - c + d & a_{22} - c + d & a_{21} - c & a_{22} - c \\ a_{11} + d & a_{12} + d & a_{11} & a_{12} \\ a_{11} + d & a_{12} + d & a_{11} & a_{22} \\ a_{11} + d & a_{12} + d & a_{21} & a_{12} \\ a_{11} + d & a_{12} + d & a_{21} & a_{22} \\ a_{21} + d & a_{22} + d & a_{11} & a_{12} \\ a_{21} + d & a_{22} + d & a_{11} & a_{22} \\ a_{21} + d & a_{22} + d & a_{21} & a_{21} \\ a_{21} + d & a_{22} + d & a_{21} & a_{22} \end{array}\right] \end{array}$$

For example, if Blue uses strategy $(1; 1, 1, 2)$ and Red uses strategy $(1; 2)$ the payoff $a_{12} - c + d$ is computed as follows: Blue reconnoiters at a cost c but receives no information about the defender, and so he selects strategy ①. Red takes countermeasures at a cost d to him or a gain of d by Blue, and uses strategy $\boxed{2}$. Therefore the payoff to Blue is $a_{12} + d - c$.

This sixteen-rowed matrix may be reduced to a four-rowed matrix by testing for dominance. Every odd row is dominated by the next even row, which eliminates eight rows. Among the even rows every $(2k + 2)$th row is dominated by the $(2k)$th row where $k = 1, 3, 5, 7$. This eliminates four more rows, leaving the following 4×4 matrix:

$$B = \begin{matrix} (1;1,1,2) \\ (1;2,1,2) \\ (2;1,1,2) \\ (2;2,1,2) \end{matrix} \overset{\begin{matrix} (1,1) & (1,2) & (2,1) & (2,2) \end{matrix}}{\begin{bmatrix} a_{11} - c + d & a_{12} - c + d & a_{11} - c & a_{22} - c \\ a_{21} - c + d & a_{22} - c + d & a_{11} - c & a_{22} - c \\ a_{11} + d & a_{12} + d & a_{11} & a_{22} \\ a_{21} + d & a_{22} + d & a_{21} & a_{22} \end{bmatrix}}.$$

It is evident from the matrix B that the decision by the attacker of whether or not to reconnoiter and by the defender of whether or not to invoke countermeasures will depend on the costs c and d. A solution of game B, i.e., the computation of a pair of optimum strategies, will indicate the advisability of reconnoitering and the use of countermeasures.

The reconnaissance problem we have described is relatively simple. It can be expanded to include the situation where the attacker receives partial information about his enemy, or where the attacker uses several types of reconnaissance. The same approach can be used to compare different types of reconnaissance in order to select the best type.

Returning to the original problem, let us solve a particular reconnaissance game. Suppose the matrix A is given by

$$A = \begin{matrix} ① \\ ② \end{matrix} \overset{\begin{matrix} \boxed{1} & \boxed{2} \end{matrix}}{\begin{bmatrix} 48 & 24 \\ 12 & 36 \end{bmatrix}},$$

and let $c = 9$, $d = 7$. Then the matrix B becomes

$$B = \begin{matrix} (1;1,1,2) \\ (1;2,1,2) \\ (2;1,1,2) \\ (2;2,1,2) \end{matrix} \overset{\begin{matrix} (1,1) & (1,2) & (2,1) & (2,2) \end{matrix}}{\begin{bmatrix} 46 & 22 & 39 & 27 \\ 10 & 34 & 39 & 27 \\ 55 & 31 & 48 & 24 \\ 19 & 43 & 12 & 36 \end{bmatrix}}.$$

This game has no saddle-point. Since no row (column) dominates any other row (column), it is necessary to solve a 4×4 game. This can be done in several ways. However, the reader can verify that the following are solutions of the game:

(a) For Blue, the attacker,

$$x_1' = (\tfrac{7}{12}, \tfrac{1}{12}, 0, \tfrac{4}{12}),$$

$$x_2' = (\tfrac{17}{36}, \tfrac{7}{36}, 0, \tfrac{12}{36}),$$

$$x_3' = (0, \tfrac{28}{144}, \tfrac{51}{144}, \tfrac{65}{144}),$$

$$x_4' = (\tfrac{28}{48}, 0, \tfrac{3}{48}, \tfrac{17}{48}).$$

(b) For Red, the defender,

$$Y' = (0, 0, \tfrac{1}{4}, \tfrac{3}{4}).$$

The value of the game is 30.

Thus the defender never takes countermeasures and defends $\frac{1}{4}$ of the time with his entire force and $\frac{3}{4}$ of the time with part of his force. The attacker has many optimal possibilities open to him, given by his large number of optimal strategies. Among his possibilities is the following: reconnoiter with probability $\frac{2}{3}$ and not reconnoiter with probability $\frac{1}{3}$. If he does reconnoiter and receives no information, then with probability $\frac{7}{8}$ he uses strategy ① and with probability $\frac{1}{8}$ he uses strategy ②; if he does reconnoiter and receives information that Red is using strategy $\boxed{1}$, then Blue selects strategy ① with probability $\frac{7}{8}$ and strategy ② with probability $\frac{1}{8}$, and similarly for strategy $\boxed{2}$. If Blue does not reconnoiter, he always uses strategy ②, or leaves a rear guard.

16. APPLICATION OF STRUCTURE THEOREMS TO RECONNAISSANCE

In the particular reconnaissance game solved in the previous section, we listed four basic solutions for Blue and one for Red. Every linear combination of these four basic solutions will be a solution of the game for Blue. We wish to verify that there are no other solutions.

Since the matrix B of the reconnaissance game is 4×4 we can represent each mixed strategy for player I by a point in a tetrahedron, $S(I)$. The four basic solutions X_1, X_2, X_3, and X_4 can be shown to lie on a plane. Taking every linear combination of these four basic solutions we obtain $X(B)$, the set of all solutions. $X(B)$ will be a plane in the tetrahedron.

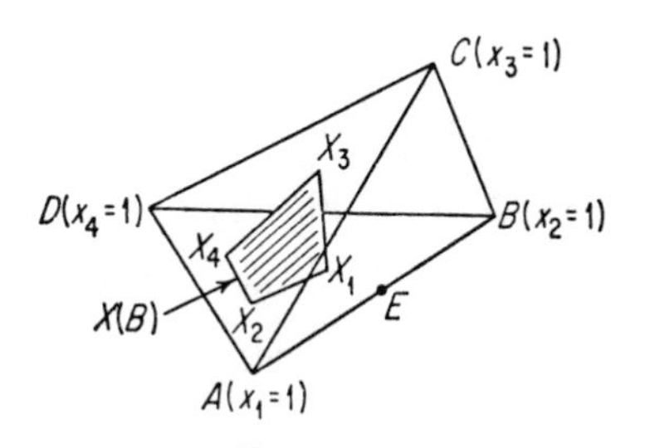

FIGURE 7

In the diagram (Fig. 7), $S(I)$ is the tetrahedron whose vertices are A, B, C, D. The four triangles are the four faces of the tetrahedron. $X(B)$ is the quadrilateral whose vertices are X_1, X_2, X_3, X_4. The point A, for example, is defined by $x_1 = 1$ and therefore is written as $A(X_1 = 1)$. Now the smallest closed simplicial face which contains $X(B)$ is the tetrahedron, or

$$S(I_1) = S(I) \quad \text{and} \quad \dim s(I_1) = \dim S(I) = 3.$$

We can represent Red's strategies geometrically (Fig. 8). $S(J)$ will be a tetrahedron. Since Red has a unique strategy $Y = (0, 0, \frac{1}{4}, \frac{3}{4})$, the set $Y(B)$ is a point F on the line $D'C'$.

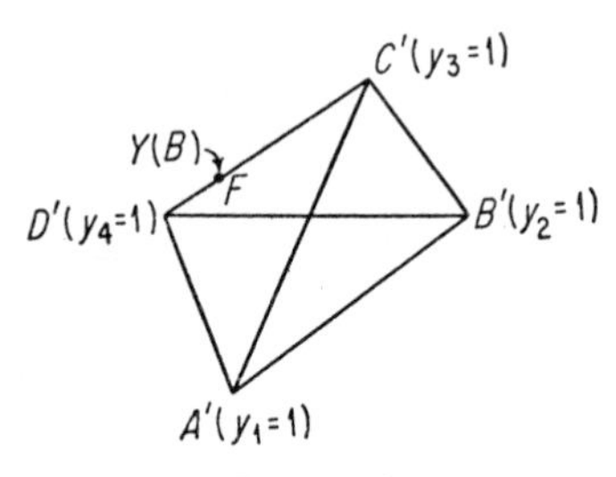

FIGURE 8

The smallest closed simplicial face containing $Y(B)$ is the line $D'C'$, or $\dim S(J_1) = 1$. Since Y is unique and its first two components are zero, it follows that the essential submatrix B_1 consists of four rows and the last two columns. The rank of this matrix is 2. Substituting in the dimensionality relationship,

$$\begin{aligned}\dim S(I_1) - \dim X(B) &= \dim S(J_1) - \dim Y(B) \\ &= \text{rank } B_1 - 1,\end{aligned}$$

we find that the relationship is satisfied, for

$$3 - 2 + 1 - 0 = 2 - 1.$$

Now

$$\begin{aligned}K_1' &= X_1'B = (34, 30, 30, 30), \\ K_2' &= X_2'B = (30, 31\tfrac{1}{3}, 30, 30), \\ K_3' &= X_3'B = (30, 37, 30, 30), \\ K_4' &= X_4'B = (37, 30, 30, 30), \\ H' &= BY \;\; = (30, 30, 30, 30).\end{aligned}$$

By definition,

$$\begin{aligned}I_1 &= \{1, 2, 3, 4\}, \\ J_1 &= \{3, 4\}.\end{aligned}$$

An examination of the above K''s and H' shows that

$$\begin{aligned}J_2 &= \{3, 4\}, \\ I_2 &= \{1, 2, 3, 4\},\end{aligned}$$

and so
$$I_1 = I_2,$$
$$J_1 = J_2.$$

If it were the case that dim $X(B) < 2$, the above verification would be sufficient to insure that $X(B)$ and $Y(B)$ are the complete sets of optimal strategies. However, dim $X(B) = 2$ in this example, and it is necessary to account for each face in $X(B)$. Now each face can be accounted for in either of the following two ways: (1) the element x_i may be zero in the two extreme points which span the face; (2) the same component k_i, in the two K's associated with the two extreme points which span the face, is equal to v and does *not* belong to J_2. Consider the reproduction of $X(B)$ (Fig. 9).

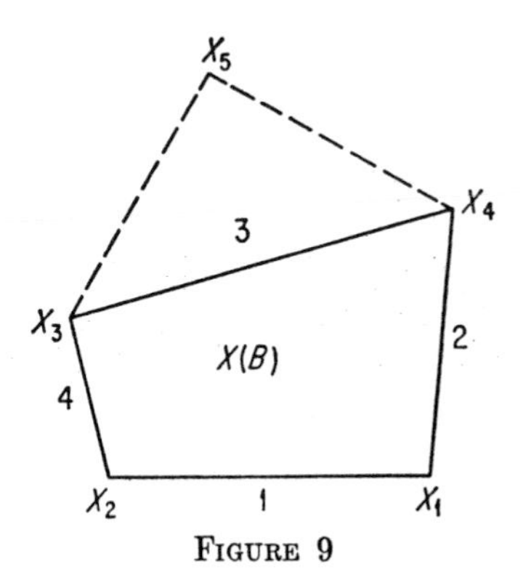

FIGURE 9

$$\begin{cases} X_4' = (\frac{28}{48}, 0, \frac{3}{48}, \frac{17}{48}) \\ K_4' = (37, 30, 30, 30) \end{cases} \qquad \begin{cases} X_3' = (0, \frac{28}{144}, \frac{51}{144}, \frac{65}{144}) \\ K_3' = (30, 37, 30, 30) \end{cases}$$

$$\begin{cases} X_1' = (\frac{7}{12}, \frac{1}{12}, 0, \frac{4}{12}) \\ K_1' = (34, 30, 30, 30) \end{cases} \qquad \begin{cases} X_2' = (\frac{17}{36}, \frac{7}{36}, 0, \frac{12}{36}) \\ K_2' = (30, 31\frac{1}{3}, 30, 30) \end{cases}$$

All of the faces except the face numbered 3 are accounted for by this process. For example, face number 1 is accounted for by the fact that $x_3 = 0$ in both X_1' and X_2'. The fact that face number 3 is not accounted for means there exists at least one more extreme point between these two. To find this extreme point glance at the sketch of $X(B)$ and its relation to $S(I)$. It is reasonable to believe that there exists only one additional extreme point and that it occurs at the intersection of the plane defined by $X(B)$ and the line DC of the tetrahedron. One of the reasons for so believing is that the existence of such an extreme point would not disturb the previous satisfaction of the first two conditions. Such a point then, if it exists, must be of the form

$$X_5' = \alpha D + (1 - \alpha)C, \qquad 0 < \alpha < 1,$$

where

$$D = (0, 0, 0, 1),$$

$$C = (0, 0, 1, 0).$$

The unknown α can be found by observing that, since $J_1 = \{3, 4\}$, the third and fourth components of $X_5'B$ are v and one finds

$$\alpha = \tfrac{1}{2}.$$

Hence

$$X_5' = (0, 0, \tfrac{1}{2}, \tfrac{1}{2}).$$

The two new faces created by the discovery of X_5' are readily explained.

The last extreme point has the property that if Blue uses this as an optimal strategy, it is uniformly better than any other strategy in $X(B)$ in that it guarantees Blue a higher expectation than any other strategy in $X(B)$ if Red deviates from his optimal strategy.

17. ATTACK ON HIDDEN-OBJECT

An interesting application of games of strategy is to the general problem of destroying a hidden-object. A defender has an object of great value which he may conceal in any one of many containers. The attacker makes a series of attempts to destroy the object by destroying the containers. Which container shall the defender use to hide the object and which containers shall the attacker attempt to destroy?

An example of this problem is the following tactical situation: A bomb is to be carried in one of two identical bombers, called P (protected) and F (flank). The bombers fly in formation, say one behind the other, so that a hostile fighter, wishing to attack the leader, must pass through the field of fire of the trailer and run the risk, α, of being shot down before being able to close in on its target. Once it has engaged its target, however, the fighter can destroy it with probability β. Since each attacking fighter has an opportunity for just one pass, its ultimate survival is of no concern. The fate of the bomb is supposed to be more important, disproportionately, than the fates of any of the aircraft.

The strategies of the two players are easily described. The defender, Red, chooses P or F, where he locates the valuable object. The attacker, Blue, chooses an ordered set of n members, such as

$$\text{F, P, F, F, \ldots, P,}$$

representing his target choice on each pass, provided the previous passes have all failed. It is agreed that if one attack has succeeded, the subsequent tries are to be directed against the other container either until it, too, has been destroyed or until Blue has used up his allotment of n tries.

If F is intact, it is harder to destroy P than if F is gone. If F is destroyed, the probability of destroying P is β. If F is intact, the probability of destroying P is $(1 - \alpha)\beta = \gamma$.

Let the payoff to the attacker be the expectation of destroying the correct container; then the game is zero-sum.

One fighter. If the attacker is permitted only one attack, then he has two strategies—attack P and attack F. The defender also has two strategies—conceal the valued object in P and conceal the valued object in F. The payoff for each outcome is given by the following 2×2 matrix.

HIDDEN-OBJECT PAYOFF

$$\begin{array}{r} \\ \\ \textit{Attack } P \\ \textit{Attack } F \end{array} \begin{array}{cc} \textit{Conceal} & \textit{Conceal} \\ \textit{in } P & \textit{in } F \\ \left[\begin{array}{c} \gamma \\ 0 \end{array}\right. & \left.\begin{array}{c} 0 \\ \beta \end{array}\right] \end{array}$$

Solving this game, we obtain that the value to the attacker is

$$v = \frac{\beta\gamma}{\beta + \gamma}.$$

The optimal strategies are always mixed, and are the same for the two players:

$$\begin{cases} \text{play P with probability } \dfrac{\beta}{\beta + \gamma} \\ \text{play F with probability } \dfrac{\gamma}{\beta + \gamma}. \end{cases}$$

Two fighters. If the attacker is permitted two attacks, he has four strategies and the defender has two strategies. The payoff matrix is obtained by computing the compound probabilities and is given by

$$\begin{array}{c} \\ P, P \\ P, F \\ F, P \\ F, F \end{array} \begin{array}{c} \begin{array}{cc} P \qquad\qquad & F \end{array} \\ \begin{bmatrix} 2\gamma - \gamma^2 & \beta\gamma \\ \gamma & \beta \\ \beta^2 - \beta\gamma + \gamma & \beta \\ \beta^2 & 2\beta - \beta^2 \end{bmatrix}. \end{array}$$

To solve this 4×2 game we note that for $0 < \gamma < \beta < 1$, $\gamma < \beta^2 - \beta\gamma + \gamma^2$. Thus the second row is dominated by the third, and may be eliminated. A further examination of some inequality relationships between β and γ will eliminate the fourth row, leaving a 2×2 game to be solved. The solution is as follows:

(i) If $\beta^2 - \beta\gamma + \gamma^2 - \gamma \geq 0$, the optimal strategies are, for the attacker,

$$\begin{cases} \text{attack P, P with probability } \dfrac{\gamma(1-\beta)+\gamma(1-\gamma)}{\beta(1-\beta)+\gamma(1-\gamma)} \\ \text{attack F, P with probability } \dfrac{\beta(1-\beta)-\gamma(1-\beta)}{\beta(1-\beta)+\gamma(1-\gamma)}; \end{cases}$$

and for the defender,

$$\begin{cases} \text{conceal in P with probability } \dfrac{\beta(1-\gamma)}{\beta(1-\beta)+\gamma(1-\gamma)} \\ \text{conceal in F with probability } \dfrac{\beta(\gamma-\beta)+\gamma(1-\gamma)}{\beta(1-\beta)+\gamma(1-\gamma)}. \end{cases}$$

The value to the attacker is

$$v = \beta\gamma\,\frac{\beta(\gamma-\beta)+2(1-\gamma)}{\beta(1-\beta)+\gamma(1-\gamma)}.$$

(ii) If $\beta^2 - \beta\gamma + \gamma^2 - \gamma \leq 0$, the optimal strategies are, for the attacker,

attack F, P;

and for the defender,

conceal in P.

The value of the game is

$$v = \beta^2 - \beta\gamma + \gamma.$$

n fighters. A typical strategy for the attacker in this case is an ordered set such as P, F, F, . . . , P, representing his target choice on each of his n passes, provided the previous passes have all failed. First, we note that any strategy in which F appears exactly r times, where $0 \leq r \leq n$, is dominated by the strategy

$$F^r = F, F, \ldots, F, P, P, \ldots, P,$$

with r F's followed by $n - r$ P's. This means that if the attacker plans to make r attacks on F, it is best to make them on the first r passes. This is the reason for the elimination of row 2 in the 4×2 matrix of the two-fighter case. It corresponded to the strategy P, F.

We have then, strategies F^r, $r = 0, 1, \ldots, n$ for the attacker and P and F for the defender. The elements $a(r, P)$ and $a(r, F)$ of the $(n + 1) \times 2$ payoff matrix are found by considering the ways the attacker can fail to destroy the valuable target. Thus:

$$\begin{aligned} a(r, P) &= 1 - (1-\beta)^r(1-\gamma)^r - r\beta(1-\beta)^{n-1}, \\ a(r, F) &= 1 - (1-\beta)^r(1-\gamma)^{n-r} \\ &\qquad - \gamma(\beta-\gamma)^{-1}(1-\beta)^r\{(1-\gamma)^{n-r} - (1-\beta)^{n-r}\}. \end{aligned}$$

The game is solved by making r continuous and making use of the concavity in r (i.e., the second derivatives are negative).

Let r_0 be the solution of

$$\frac{da(r, \mathrm{P})}{dr} = 0,$$

and let $R = \frac{1 - \gamma}{1 - \beta}$. Then

$$r_0 = n - \frac{\ln \beta/(1 - \beta) - \ln \ln \mathrm{P}}{\ln R}.$$

The solution depends on the value of r_0. If $a(r_0, \mathrm{P}) \leq a(r_0, \mathrm{F})$, then the attacker has a pure optimal strategy r_0 and the defender also has a pure optimal strategy P. If $a(r_0, \mathrm{P}) \geq a(r_0, \mathrm{F})$, then the attacker has a pure optimal strategy r given by the solution of the equation

$$\frac{r\beta(\beta - \gamma)}{\gamma(1 - \beta)} = \left(\frac{1 - \gamma}{1 - \beta}\right)^{n-r},$$

or approximately $r = n\gamma/(\beta + \gamma)$. The defender has an optimal strategy which is a mixture of P and F.

For the original discrete game, define the integer $r_1 = [r_0]$ and solve the 2×2 matrix

$$\begin{bmatrix} a(r_1, \mathrm{P}) & a(r_1, \mathrm{F}) \\ a(r_1 + 1, \mathrm{P}) & a(r_1 + 1, \mathrm{F}) \end{bmatrix}.$$

If $a(r_1, \mathrm{P}) \leq a(r_1, \mathrm{F})$, we have pure strategy solutions. If $a(r_1, \mathrm{P}) \geq a(r_1, \mathrm{F})$, we have mixed strategies solutions.

18. SELECTING A PARTICULAR OPTIMAL STRATEGY

If a player has many optimal strategies, he may use any one of them and he can expect to receive at least an amount v, the value of the game, regardless of the strategy used by his opponent. If his opponent also uses an optimal strategy, then the player can expect to receive exactly this amount v. Thus if both players play optimally, there is no reason for preferring one optimal strategy over another except for reasons extraneous to the game model. (Examples of extraneous considerations are preferences for pure strategies or preferences for mixed strategies having simple components.) However, if we permit the unlikely possibility that an opponent may make some mistake and fail to choose one of his optimal strategies then, by selecting a particular optimal strategy, a player can take maximum advantage of the mistake.

Among the solutions for each player there exist certain strategies which, in addition to being optimal, take maximum advantage of the mistakes, if any, of one's opponent. We shall describe one method for choosing such an optimal strategy. This particular optimal strategy will be chosen on the assumption that the opponent will make some mistake, unknown in advance to the player, but that the opponent will try to minimize the result of such mistake.

Let us assume that Blue, the maximizing player, wishes to select an optimal strategy which takes maximum advantage of Red's possible departures from his set of optimal strategies. Let the payoff matrix be A and the set of optimal strategies of Blue be $T_1(A)$. A procedure for selecting a particular optimal strategy is as follows:

(i) Assume that Blue uses only mixed strategies belonging to $T_1(A)$. This is equivalent to replacing the maximizing player's pure strategies by the extreme points of $T_1(A)$.

(ii) Assume that Red uses only those pure strategies which yield more than the value of the game to Blue for at least one optimal strategy of Blue. This is equivalent to deleting from Red's pure strategies all pure strategies which yield the value of the game against every optimal strategy of Blue.

(iii) Using the matrix A, compute the new payoff matrix $\bar{A}$ associated with the set of strategies defined in (i) and (ii).

(iv) Find the optimal strategies $T_1(\bar{A})$ of Blue. It is evident that $T_1(\bar{A})$ is contained in $T_1(A)$. If Blue uses any strategy in $T_1(\bar{A})$, he will maximize the minimum gain for any departure of Red from his optimal strategies.

(v) If $T_1(\bar{A})$ is not unique, the above process may be repeated for $\bar{A}$ and $T_1(\bar{A})$. The process will either terminate with a single strategy for Blue or every element of the payoff matrix, $\bar{\bar{A}}$ derived from $\bar{A}$ will be identical, in which case all strategies of $T_1(\bar{\bar{A}})$ are equivalent—they all take maximum advantage of mistakes.

Example. In the game of Morra described in section 3 of Chapter 1, we had the following 9×9 payoff matrix:

$$A = \begin{bmatrix} 0 & 2 & 2 & -3 & 0 & 0 & -4 & 0 & 0 \\ -2 & 0 & 0 & 0 & 3 & 3 & -4 & 0 & 0 \\ -2 & 0 & 0 & -3 & 0 & 0 & 0 & 4 & 4 \\ 3 & 0 & 3 & 0 & -4 & 0 & 0 & -5 & 0 \\ 0 & -3 & 0 & 4 & 0 & 4 & 0 & -5 & 0 \\ 0 & -3 & 0 & 0 & -4 & 0 & 5 & 0 & 5 \\ 4 & 4 & 0 & 0 & 0 & -5 & 0 & 0 & -6 \\ 0 & 0 & -4 & 5 & 5 & 0 & 0 & 0 & -6 \\ 0 & 0 & -4 & 0 & 0 & -5 & 6 & 6 & 0 \end{bmatrix}.$$

The value of this game is zero. It can be verified that each player's set of solutions has the following four extreme points:

$$(a) = (0,\ 0,\ \tfrac{5}{12},\ 0,\ \tfrac{4}{12},\ 0,\ \tfrac{3}{12},\ 0,\ 0),$$

$$(b) = (0,\ 0,\ \tfrac{16}{37},\ 0,\ \tfrac{12}{37},\ 0,\ \tfrac{9}{37},\ 0,\ 0),$$

$$(c) = (0,\ 0,\ \tfrac{20}{47},\ 0,\ \tfrac{15}{47},\ 0,\ \tfrac{12}{47},\ 0,\ 0),$$

$$(d) = (0,\ 0,\ \tfrac{25}{61},\ 0,\ \tfrac{20}{61},\ 0,\ \tfrac{16}{61},\ 0,\ 0).$$

Replacing Blue's nine pure strategies by the four extreme points, we obtain the following 4×9 matrix:

$$B = \begin{bmatrix} \frac{2}{12} & 0 & 0 & \frac{1}{12} & 0 & \frac{1}{12} & 0 & 0 & \frac{2}{12} \\ \frac{4}{37} & 0 & 0 & 0 & 0 & \frac{3}{37} & 0 & \frac{4}{37} & \frac{10}{37} \\ \frac{8}{47} & \frac{3}{47} & 0 & 0 & 0 & 0 & 0 & \frac{5}{47} & \frac{8}{47} \\ \frac{14}{61} & \frac{4}{61} & 0 & \frac{5}{61} & 0 & 0 & 0 & 0 & \frac{4}{61} \end{bmatrix}.$$

Deleting from Red's strategies all pure strategies which yield the value of the game against every optimal strategy of Blue (delete columns three, five, and seven), we obtain the 4×6 matrix:

$$\bar{A} = \begin{bmatrix} \frac{2}{12} & 0 & \frac{1}{12} & \frac{1}{12} & 0 & \frac{2}{12} \\ \frac{4}{37} & 0 & 0 & \frac{3}{37} & \frac{4}{37} & \frac{10}{37} \\ \frac{8}{47} & \frac{3}{47} & 0 & 0 & \frac{5}{47} & \frac{8}{47} \\ \frac{14}{61} & \frac{4}{61} & \frac{5}{61} & 0 & 0 & \frac{4}{61} \end{bmatrix}.$$

Solving $\bar{A}$, we find that Blue's set of solutions has two extreme points

$$(\tfrac{120}{275},\ 0,\ \tfrac{94}{275},\ \tfrac{61}{275}),$$

$$(\tfrac{12}{110},\ \tfrac{37}{110},\ 0,\ \tfrac{61}{110}).$$

In terms of the original nine strategies of Blue, these two solutions yield the same solution of A, namely:

$$(e) = (0,\ 0,\ \tfrac{23}{55},\ 0,\ \tfrac{18}{55},\ 0,\ \tfrac{14}{55},\ 0,\ 0).$$

The value of the game with payoff matrix $\bar{A}$ is $\frac{2}{55}$. Therefore if Blue uses the particular optimal strategy (e), he will gain at least $\frac{2}{55}$ if Red ever departs from an optimal strategy in the original game.

4 GAMES IN EXTENSIVE FORM

1. REPRESENTATION OF GAMES

Finite games can be conveniently represented with the help of topological trees. We shall give an example of such a representation for the following two-person game. Suppose that the first move of the game is made by Blue, who has to choose from among four alternatives. We represent this situation by the bottom part of Fig. 10, where B at the lowest

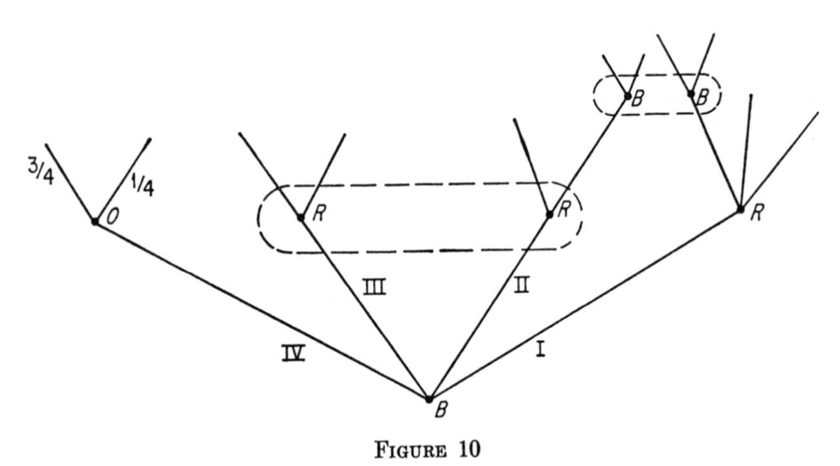

FIGURE 10

point of the figure indicates that the first move is made by Blue, and the four rising lines indicate that Blue has four alternatives at this move. Let us number the alternatives at this first move (and, indeed, at each of the succeeding moves) in a counterclockwise sense with roman numerals. Thus alternative I at the first move is the one corresponding to the segment leading to the vertex marked R in the figure.

Now suppose that, if Blue chooses I on this first move, then the next move is to be made by Red, who in turn has a choice from among three alternatives; and suppose that in addition it is given that if Blue chooses II or III on the first move, then the next move is made by Red, who in either case is to choose from two alternatives. This is indicated by putting an R at the end of the segment corresponding to the first choice by Blue, drawing three lines upward from this point, putting an R on the vertices corresponding to moves II and III of Blue, and drawing two lines upward from each of these points.

Suppose, moreover, that if Blue chooses IV on the first move, then the next move is made by a chance device which chooses from two alternatives. Suppose that this chance device is such that it assigns probability $\frac{1}{4}$ to the first chance alternative, and $\frac{3}{4}$ to the second. We represent this in Fig. 10 by putting an O at the vertex corresponding to the move made by chance; and putting $\frac{1}{4}$ and $\frac{3}{4}$ on the appropriate lines rising from this vertex.

Finally, let us suppose that if Blue chooses I and Red chooses III, or if Blue chooses II and Red chooses I, then the next move is to be made by Blue again; that he then has a choice of two alternatives; and that these are the only possible moves in the game. Then the tree in Fig. 10 represents the complete structure of the moves of the game.

Any unicursal path from the bottom to the top of the tree represents a play of the game. There are just as many plays as there are top points of the tree. Thus there are precisely 11 possible ways of playing the game represented in Fig. 10. For each of these ways of playing the game there is a payoff to each player.

In order to complete the description of the game, it is also necessary to specify how much the players know about the previous choices at the time they make their moves. Thus suppose that when Red makes his choice he does not know whether Blue has chosen II or III on his first move; we can indicate this, as in Fig. 10, by enclosing the two corresponding points in a region bounded by a broken line. We call such a set of vertices among which a player cannot distinguish when he makes his move an *information set.* If a player at a certain point is completely informed about the past course of the play, the corresponding information set contains only the single point, in which case we omit the broken lines.

If each information set contains just one vertex, we say that the game is one with *perfect information;* that is, each player, at the time he makes each of his moves, is completely informed about the past course of the play.

In general, a finite game can be represented as above by a finite tree. The partition of the vertices into information sets is subject to the obvious restrictions that all the vertices in a given information set must correspond to the same player, and must present the same number of alternatives. In addition, we require that no information set intersect any play in more than one point.

A strategy for a player will be a function which is defined over the class of his information sets, and for each information set picks out one of the available alternatives. Thus, for the game represented in Fig. 10, let α be the set consisting of the bottom vertex and let β be the set consisting of the two top vertices; then a strategy for Blue is a function f such that

$$f(\alpha) \in \{\text{I, II, III, IV}\} \quad \text{and} \quad f(\beta) \in \{\text{I, II}\}.$$

Thus Blue has six strategies in the game. For example, the strategy (II, I) says that Blue first picks his second alternative and on his second move he picks his first alternative.

2. GAMES WITH PERFECT INFORMATION—SADDLE-POINTS

We shall now examine a special class of games—the so-called games with "perfect information." These games have the property that at every point in every play each player whose turn it is to move knows exactly what choices have been made previously. The moves are made alternately, which means that, in terms of the graph, each information set consists of one element.

Many common parlor games are games with perfect information. Ticktacktoe, checkers, chess, backgammon, for instance, are games with perfect information.

We shall show that the payoff matrix of any zero-sum two-person game with perfect information has a saddle-point—i.e., that there are optimal pure strategies for such a game.

In order to carry out the proof, it is convenient to introduce the notion of the *truncation* of a game with perfect information. Truncations are those games which arise from a given game if the first move is deleted. The number of truncations is the number of alternatives available at the first move. The payoff functions for these truncated games are the original payoff functions, with their domains of definition suitably restricted. We can also consider truncation of a strategy—it picks out the same alternatives at the branch points as does the original strategy.

We shall prove that a perfect-information game has a saddle-point by an induction on the length of the game—i.e., on the number of branch points in the longest possible play of the game.

For games of length 1 (i.e., games with one move) the theorem is obvious. Now suppose that the theorem is true for all games of length less than k. Let Γ be a game of length k. Suppose there are r alternatives on the first move, and let $\Gamma_1, \Gamma_2, \ldots, \Gamma_r$ be the r truncations of Γ. Let $A_i^{(B)}$ be

Blue's set of pure strategies for the game Γ_i. Let $A^{(B)}$ and $A^{(R)}$ be Blue's and Red's set of pure strategies, respectively, in the game Γ.

By the induction hypothesis, there is a saddle-point in each of the games Γ_i. For each i, let f_i^*, g_i^* be a saddle-point. Let M_i be the payoff function in the game Γ_i, then we have

$$\begin{cases} M_i(f_i^*, g_i^*) \geq M_i(f_i, g_i^*) \\ M_i(f_i^*, g_i^*) \leq M_i(f_i^*, g_i) \end{cases} \tag{4.1}$$

for $i = 1, 2, \ldots, r$ and for f_i any member of $A_i^{(B)}$ and g_i any member of $A_i^{(R)}$.

We need to distinguish three cases according to who makes the first move—Red, Blue, or chance.

Case 1. The first move of Γ is a chance move. Letting q be a branch point of one of the truncated games Γ_i, and letting it correspond to a move made by Blue, we set

$$f^*(q) = f_i^*(q).$$

If q corresponds to a move of Red, then set

$$g^*(q) = g_i^*(q).$$

Since the first move is made by chance, it is clear that f^* is defined over every branch of Γ which corresponds to a move made by Blue, and hence is a member of $A^{(B)}$. We shall show that f^*, g^* is a saddle-point of Γ.

Let the probabilities assigned to the r alternatives at the first move be $\alpha_1, \alpha_2, \ldots, \alpha_r$. Let M be the payoff function in the game Γ. Then, if f and g are strategies in Γ,

$$M(f, g) = \Sigma\, \alpha_i M_i(f_i, g_i).$$

In particular, since $f_1^*, \ldots, f_r^*$ are truncations of f^* and $g_1^*, \ldots, g_r^*$ are truncations of g^*, we have

$$\begin{cases} M(f, g^*) = \Sigma\, \alpha_i M_i(f_i, g_i^*) \\ M(f^*, g) = \Sigma\, \alpha_i M_i(f_i^*, g_i) \end{cases} \tag{4.2}$$

and

$$M(f^*, g^*) = \Sigma\, \alpha_i M_i(f_i^*, g_i^*). \tag{4.3}$$

From (4.1) and (4.2) it follows that

$$\Sigma\, \alpha_i M_i(f_i, g_i^*) \leq \Sigma\, \alpha_i M_i(f_i^*, g_i^*) = M(f^*, g^*),$$
$$\Sigma\, \alpha_i M_i(f_i^*, g_i) \geq \Sigma\, \alpha_i M_i(f_i^*, g_i^*) = M(f^*, g^*).$$

Hence (f^*, g^*) is a saddle-point of Γ.

Case 2. The first move, q_0, of Γ is made by Blue. Let

$$\max_{i \leq r} M_i(f_i^*, g_i^*) = M_\mu(f_\mu^*, g_\mu^*). \tag{4.4}$$

Now define a function f^* by setting

$$(4.5) \qquad f^*(q_0) = \mu \quad \text{and} \quad f^*(q) = f_i^*(q)$$

for any point q in the truncated games Γ_i which correspond to a move made by Blue. We define g^* in the same way as in the previous case.

It is clear that f^* and g^* are strategies in Γ. We shall show that they yield a saddle-point of Γ.

If g is any strategy for Red in Γ, and if g_μ is its truncation to Γ_μ, then

$$M(f^*, g) = M_\mu(f_\mu^*, g_\mu).$$

In particular, we have

$$(4.6) \qquad M(f^*, g^*) = M_\mu(f_\mu^*, g_\mu^*).$$

Thus, if g is any strategy for Red in Γ, and if g_μ is its truncation to Γ_μ, we see that

$$M(f^*, g^*) = M_\mu(f_\mu^*, g_\mu^*) \leq M_\mu(f_\mu^*, g_\mu) = M(f^*, g).$$

Now let f be any strategy for Blue in Γ, and suppose that f picks out the ith alternative on the first move, i.e., suppose that

$$f(q_0) = i.$$

Let f_i be the truncation of f to Γ_i. Then we see that, if g is any strategy for Red in Γ, and if g_i is its truncation to Γ_i,

$$M(f, g) = M_i(f_i, g_i).$$

In particular,

$$M(f, g^*) = M_i(f_i, g_i^*).$$

But, from (4.4) we have

$$M_\mu(f_\mu^*, g_\mu^*) \geq M_i(f_i^*, g_i^*).$$

Hence, we can conclude from (4.6) that

$$M(f^*, g^*) = M_\mu(f_\mu^*, g_\mu^*) \geq M_i(f_i^*, g_i^*) \geq M_i(f_i, g_i^*) = M(f, g^*).$$

We see that (f^*, g^*) is a saddle-point as was to be shown.

Case 3. The first move of Γ is made by Red. This is analogous to the preceding case.

5 METHODS OF SOLVING GAMES

1. SOLVING FOR OPTIMAL STRATEGIES

From the fundamental theorem of game theory, the minimax theorem, it follows that each player has an optimal strategy. Using an optimal strategy (generally a mixed strategy), a player can expect to win a fixed amount (the game value) regardless of the strategy selected by his opponent, and this fixed amount is as large as is strategically possible. Of course, he may win more than this fixed amount from his opponent if his opponent does not use an optimal strategy. In order to protect himself against this unnecessary loss, the opponent will also use an optimal strategy. We have referred to a pair of optimal strategies, one for each player, as a *solution of the game.*

The determination of all solutions of a game—or of even a single solution—is generally a lengthy computing task. As might be expected, the volume of computations increases with the number of pure strategies available to the two players, or the number of elements in the payoff matrix. If at least one player has 2 or 3 pure strategies, it is possible to represent the strategies in two dimensions and then obtain a graphical solution.

We shall discuss various methods of obtaining some solution of a game. We shall also describe methods for getting all solutions. Single solutions will be obtained either by guessing a solution and then verifying it, or by successive approximations. To obtain all solutions we shall, in effect, solve a large number of linear systems of equations.

Since the volume of computing required to obtain a solution depends on the number of strategies, it is important to reduce this number whenever possible. The dominance criteria discussed in the previous chapter can be applied for this purpose. It will be recalled that elimination by nonstrict dominance may lose some solutions but elimination by strict dominance preserves all solutions.

We have already described one method for getting a solution—namely, the simplex algorithm. We shall now discuss five other methods of solving a game. They are: guess and verify, examination of submatrices, successive approximations, graphical, and mapping.

2. GUESS AND VERIFY

Given a payoff matrix with a large number of elements we first examine it for pure strategy solutions or saddle-points. This can be done very easily. In order that a game have a pure strategy solution, there must exist an element in the payoff matrix which is simultaneously the minimum of its row and the maximum of its column. If no such element exists there are no pure strategy solutions.

Example. The game defined by

$$\begin{array}{l c c} & & \text{Min of row} \\ & \begin{bmatrix} 3 & 7 & 4 & 3 \\ 13 & 10 & 7 & 8 \\ 10 & 4 & 1 & 9 \\ 3 & 5 & 6 & 7 \end{bmatrix} & \begin{array}{c} 3 \\ \underline{7} \\ 1 \\ 3 \end{array} \\ \text{Max of column} & \begin{array}{cccc} 13 & 10 & \underline{7} & 9 \end{array} & \end{array}$$

has a pure strategy solution—Blue's second strategy and Red's third strategy—and the value of the game is 7, since this element is the minimum of row 2 and the maximum of column 3.

If the game does not have a pure strategy solution, i.e., a saddle-point, and therefore has a mixed strategy solution, it may be possible to guess a solution or the form of a solution. Some idea of the form of the solution—e.g., the pure strategies which must be in the solution—can be obtained from the particular game being solved or from the elements in the payoff matrix.

If an analysis of the background of the game provides a guess of a solution of the game, we can verify it very readily. In order for X_0, Y_0 to be a solution, they must satisfy

$$\min_Y X_0'AY = \max_X X'AY_0 = X_0'AY_0.$$

Example. Suppose that the payoff matrix is given by

$$\begin{bmatrix} 1 & -1 & -5 \\ -1 & 1 & -2 \\ -5 & 2 & 0 \end{bmatrix}.$$

We guess that each player wishes to avoid using his third strategy, since that strategy can cause a loss of as much as 5 but a gain of at most 2. We therefore guess that $x_3 = y_3 = 0$ and then we need consider only the smaller game defined by the 2×2 matrix,

$$\begin{bmatrix} 1 & -1 \\ -1 & 1 \end{bmatrix}.$$

It is obvious that this smaller game has for its solution $x_1 = x_2 = \frac{1}{2}$, $y_1 = y_2 = \frac{1}{2}$. Since for the original game we have

$$\min_Y X_0'AY = \max_X X'AY_0 = 0,$$

it follows that we have arrived at a solution to the game.

3. EXAMINATION OF SUBMATRICES

The method of examination of submatrices, as discussed in Chapter 3, provides all the solutions of a game by obtaining certain basic solutions and then expressing every solution as a linear combination of these basic solutions. The method consists of testing each square submatrix of the payoff matrix for all strategies active, and then testing an expanded vector for a solution of the full matrix.

Let $A = (a_{ij})$ be the payoff matrix having m rows and n columns. Consider some $r \times r$ submatrix $B = (b_{ij})$ of A, where $r = 1, 2, \ldots,$ $\min(m, n)$. Let us test whether this subgame has all strategies active (i.e., whether there exists some X_0, Y_0 such that $X_0'BY = X'BY_0 = X_0'BY_0$ for all X and Y). This can be done (as was shown in Chapter 3) by constructing the adjoint matrix $J = (B_{ji})$ of the matrix B and then computing all the row-sums $\sum_j B_{ji}$ and column-sums $\sum_i B_{ji}$. If all $2r$ sums are non-negative, or all nonpositive, then the pair of r-component vectors

$$x_i = \frac{\sum_j B_{ji}}{\sum_i \sum_j B_{ji}}, \qquad y_j = \frac{\sum_i B_{ji}}{\sum_i \sum_j B_{ji}},$$

provide a solution of the subgame B.

It is next necessary to test whether the solution of the subgame B is a solution of the game A. Introduce zero components for the $m - r$ rows and $n - r$ columns of the matrix A which did not appear in submatrix B. This yields the full vectors $X = (x_i)$, $Y = (y_j)$. Test whether the full vectors X and Y solve A—that is, whether

$$\max_i \sum_{j=1}^{n} a_{ij}y_j = \min_j \sum_{i=1}^{m} a_{ij}x_i.$$

If X, Y solves A then it is said to be a *basic solution* of A and the value of the game is

$$v = \max_i \sum_j a_{ij}y_j = \frac{|B|}{\sum_i \sum_j B_{ji}} = \min_j \sum_{i=1}^{m} a_{ij}x_i.$$

If this procedure is applied to every square submatrix of A we shall obtain all the basic solutions of A. Every basic solution of A is produced by some square submatrix of A. Since there are a finite number of submatrices of A there will be a finite number of basic solutions.

All solutions of A are obtained by taking all possible convex linear combinations of the basic solutions. Thus if $X_1, X_2, \ldots, X_l$ are the l basic solutions of Blue, then all his solutions are given by

$$\alpha_1 X_1 + \alpha_2 X_2 + \ldots + \alpha_l X_l,$$

where $\alpha_i \geq 0$ and $\sum_{i=1}^{l} \alpha_i = 1$.

4. SUCCESSIVE APPROXIMATIONS

It is possible to obtain the value and a solution of a given game by a relatively simple method of successive approximations. The method requires only two operations: location of the maximum or minimum of a discrete set of numbers, and addition.

Given a game defined by a payoff matrix $A = (a_{ij})$, whose solution is unknown, then one way of determining an optimal strategy is to play the game many times, each time selecting that pure strategy which is best against the opponent's total performance to that play. The relative frequencies of these strategies will yield an approximate solution to the game.

The method can best be illustrated by an example. Suppose we are given the game defined by the payoff matrix

$$\begin{array}{c} \\ ① \\ ② \\ ③ \end{array} \begin{array}{c} \begin{array}{ccc} \boxed{1} & \boxed{2} & \boxed{3} \end{array} \\ \begin{bmatrix} 2 & 1 & 0 \\ 2 & 0 & 3 \\ -1 & 3 & -3 \end{bmatrix} \end{array},$$

where the numbers in circles represent Blue's strategies and those in squares, Red's strategies. Assuming that Blue begins the series of plays by selecting ①, the successive approximations are shown by Table 1. The symbols in the table have the following meaning: N is the number of the play; $i(N)$ is the pure strategy chosen by Blue for the Nth play; $K_1(N)$ is the total receipts of Blue after N of his plays if Red used his pure strategy $\boxed{1}$ constantly, and similarly $K_2(N)$ and $K_3(N)$; $\underline{v}(N)$ is the least that Blue

Table 1. Successive Approximations

N	$i(N)$	$K_1(N)$	$K_2(N)$	$K_3(N)$	$\underline{v}(N)$	$j(N)$	$H_1(N)$	$H_2(N)$	$H_3(N)$	$\bar{v}(N)$	$\bar{v}(N) - \underline{v}(N)$
1	①	2	1	0	.000	$\boxed{3}$	0	$\underline{3}$	−3	3.000	3.000
2	②	4	$\underline{1}$	3	.500	$\boxed{2}$	1	$\underline{3}$	0	1.500	1.000
3	②	6	$\underline{1}$	6	.333	$\boxed{2}$	2	$\underline{3}$	3	1.000	.667
4	②	8	$\underline{1}$	9	.250	$\boxed{2}$	3	3	$\underline{6}$	1.500	1.250
5	③	7	$\underline{4}$	6	.800	$\boxed{2}$	4	3	$\underline{9}$	1.800	1.000
6	③	6	7	$\underline{3}$	.500	$\boxed{3}$	4	$\underline{6}$	6	1.000	.500
7	②	8	7	$\underline{6}$	.857	$\boxed{3}$	4	$\underline{9}$	3	1.286	.429
8	②	10	$\underline{7}$	9	.875	$\boxed{2}$	5	$\underline{9}$	6	1.125	.250
9	②	12	$\underline{7}$	12	.778	$\boxed{2}$	6	$\underline{9}$	9	1.000	.222
10	②	14	$\underline{7}$	15	.700	$\boxed{2}$	7	9	$\underline{12}$	1.200	.500
11	③	13	10	12	.909	$\boxed{2}$	8	9	$\underline{15}$	1.364	.455
12	③	12	13	$\underline{9}$	.750	$\boxed{3}$	8	$\underline{12}$	12	1.000	.250

can expect to receive, on the average, after N of his plays; $j(N)$ is the pure strategy chosen by Red for his Nth play; $H_1(N)$ is the total receipts of Blue after N plays of Red against the constant strategy ① of Blue, and similarly $H_2(N)$ and $H_3(N)$; $\bar{v}(N)$ is the most that Blue can expect to receive on the average, after N plays of Red.

$$\underline{v}(N) = \frac{1}{N} \min_j K_j(N) \quad \text{and} \quad \bar{v}(N) = \frac{1}{N} \max_i H_i(N).$$

Table 1 has been completed as follows: For the first play of the game, assume that Blue chooses ①. Then Blue will receive 2, 1, or 0 depending on whether Red chooses $\boxed{1}$, $\boxed{2}$, or $\boxed{3}$. Red will therefore choose $\boxed{3}$ for his first play, since that minimizes Blue's receipts; and Blue will thus receive 0, 3, or −3. For the second play, Blue will choose ② since that maximizes his receipts against Red's first play. Thus after two plays, Blue has received a total of 4, 1, or 3 depending on whether Red chooses $\boxed{1}$, $\boxed{2}$, or $\boxed{3}$. Red will therefore choose $\boxed{2}$ since that minimizes Blue's receipts for $N = 2$, and makes Blue's receipts total 1, 3, or 0, depending on whether Blue chooses ①, ②, or ③. We obtain

$$\underline{v}(2) = \tfrac{1}{2} = 0.50 \quad \text{and} \quad \bar{v}(2) = \tfrac{3}{2} = 1.50.$$

The process is identical for all successive N.

To obtain an approximation to an optimal strategy, we determine the relative frequencies of each of the pure strategies in the table. Thus at $N = 12$, we have

$$X = (\tfrac{1}{12}, \tfrac{7}{12}, \tfrac{4}{12}), \qquad Y = (\tfrac{0}{12}, \tfrac{8}{12}, \tfrac{4}{12}).$$

The value of the game is approximated by $\underline{v}(N)$ and $\bar{v}(N)$. Thus at $N = 12$, the value of the game is between 0.75 and 1.00.

For convenience, we adopt the rule that if the maximum or minimum is assumed for more than one strategy, then we use the first of these strategies.

We might summarize the successive approximation procedure formally as follows:

(i) The procedure is started with Blue's picking strategy $i(1) = ①$ for $N = 1$, yielding $K_1(1), K_2(1), \ldots, K_n(1)$.

(ii) Successive $i(N)$ and $j(N)$ are selected as follows:

(a) $j(N)$ is picked such that it is the smallest integer for which

$$\min [K_1(N), K_2(N), \ldots, K_j(N), \\ K_{j(N)+1}(N), \ldots, K_n(N)] = K_{j(N)}(N).$$

(b) $i(N)$ is picked such that it is the smallest integer for which

$$\max [H_1(N-1), H_2(N-1), \ldots, H_{i(N)}(N-1), \\ H_{i(N)+1}(N-1, \ldots, H_m(N-1)] = H_{i(N)}(N-1).$$

(iii) Successive $H_i(N)$ and $K_j(N)$ are computed as follows:

(a) $$H_i(1) = a_{ij(1)} \qquad \text{for } N = 1,$$

(b) $$H_i(N) = H_i(N-1) + a_{ij(N)} \qquad \text{for } N > 1.$$

(iv) Approximations to the value of the game are given by

(a) $$\underline{v}(N) = \frac{1}{N} K_{j(N)}(N),$$

(b) $$\overline{v}(N) = \frac{1}{N} H_{i(N)}(N).$$

We may write $i(N)$ as a column matrix with $m - 1$ zeros and a single 1 in the position of that particular pure strategy. Thus if $i(5) = 2$, and $m = 3$, we may write

$$i(5) = \begin{bmatrix} 0 \\ 1 \\ 0 \end{bmatrix}.$$

Similarly, we may express $j(N)$ as a column matrix with $n - 1$ zeros and a single 1.

After N steps, an approximation to an optimal strategy will be

$$X(N) = \frac{1}{N} \sum_{k=1}^{N} i(k), \qquad Y(N) = \frac{1}{N} \sum_{l=1}^{N} j(l).$$

We have, for all N,

$$\overline{v}(N) \geq v \geq \underline{v}(N).$$

If

$$\lim_{N\to\infty} X(N) \quad \text{and} \quad \lim_{N\to\infty} Y(N)$$

exist, then these limits are a solution of the game, and the value is

$$v = \lim_{N\to\infty} \bar{v}(N) = \lim_{N\to\infty} \underline{v}(N).$$

In the preceding example, we have

$$\lim_{N\to\infty} X(N) = \begin{bmatrix} 0 \\ \frac{2}{3} \\ \frac{1}{3} \end{bmatrix}, \qquad \lim_{N\to\infty} Y(N) = \begin{bmatrix} 0 \\ \frac{2}{3} \\ \frac{1}{3} \end{bmatrix}, \quad \text{and} \quad v = 1.00.$$

It can be shown that $\bar{v}(N)$ and $\underline{v}(N)$ will always converge to v, the value of the game. However, the strategies $X(N)$ and $Y(N)$ may not converge. If the strategies fail to converge, the reason is generally the oscillating character of the $X(N)$ and $Y(N)$ around a solution. Making use of this fact, we still find it possible to determine an optimal strategy by this method.

5. GRAPHICAL SOLUTION OF 3 × 3 GAMES

In Chapter 2 we described a graphical method of solving 2 × 2 games—i.e., games in which each player has two strategies. The same method can be used for games in which one player has n strategies and the other player has two strategies. Since it is possible to represent a three-dimensional vector in two dimensions, we can solve graphically in a plane any game in which each player has three strategies.

If (a_1, a_2, a_3) is a three-dimensional vector it can be represented in a plane by means of the triaxial coordinates $(\alpha_1, \alpha_2, \alpha_3)$, with

$$\alpha_1 = 3a_1 - s$$

$$\alpha_2 = 3a_2 - s$$

$$\alpha_3 = 3a_3 - s$$

$$\alpha_1 + \alpha_2 + \alpha_3 = 0$$

where $s = a_1 + a_2 + a_3$, and α_i is the length of the vector projected on the ith axis.

If we set

$$\beta_i = k\alpha_i + \tfrac{1}{3}$$

where k is any convenient scale factor, we obtain the corresponding *triangular coordinates* $(\beta_1, \beta_2, \beta_3)$ for which

$$\beta_1 + \beta_2 + \beta_3 = 1.$$

Ordinary rectangular coordinates may also be used to represent a three-dimensional vector by taking, for example, the first two components (α_1, α_2).

To solve a 3×3 game, we plot the rows and columns separately on one of these systems of coordinates in two dimensions and consider the two triangles formed by the two sets of three points. If both triangles contain the origin, then:

(a) The game has a solution in which all strategies are active.
(b) The solution is unique and the game is completely mixed if and only if the origin is actually interior to both triangles.
(c) The positions of the origin with respect to triangular coordinates will determine the weights which each player should attach to his strategies. For example, in Fig. 11, plotted in triaxial coordinates, strategy ② receives a weight OB/AOB.

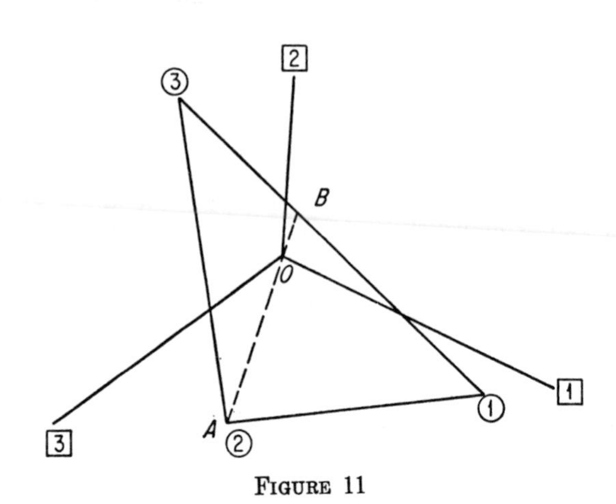

FIGURE 11

If all strategies are not active, further information may be obtained from the triangle not having the origin in its interior (possibly both triangles):

(a) Construct any line through the origin so that the triangle lies entirely in one half-plane.
(b) The vertices of the triangle correspond to the strategies of one player. Represent those of the other player by any set of three points on the positive axes of the triaxial system, for example the points

$$(2, -1, -1), (-1, 2, -1), (-1, -1, 2).$$

In triangular coordinates, use the three vertices

$$(1, 0, 0), (0, 1, 0), (0, 0, 1).$$

(c) Then some combination of the strategies which lie in the same half-plane with the triangle dominates some combination of those in the opposite half-plane. If open half-planes can be used, then the dominance is strict.

We can interpret the weak dominance to mean that the dominating row (dominating column) must appear in the solution, while at least one of the other two may be excluded.

Example. Let the payoff matrix be given by

$$\begin{array}{c} \\ ① \\ ② \\ ③ \end{array}\begin{array}{c} \begin{array}{ccc} \boxed{1} & \boxed{2} & \boxed{3} \end{array} \\ \begin{bmatrix} -1 & 3 & -3 \\ 2 & 0 & 3 \\ 2 & 1 & 0 \end{bmatrix} \end{array}.$$

The triaxial representation of the strategies are as follows:

$$\begin{aligned} ① &= (-2, 10, -8), & \boxed{1} &= (-6, 3, 3), \\ ② &= (1, -5, 4), & \boxed{2} &= (5, -4, -1), \\ ③ &= (3, 0, -3), & \boxed{3} &= (-9, 9, 0). \end{aligned}$$

Plotting the strategies on triaxial coordinates, we obtain the two triangles ① ② ③ and $\boxed{1}$ $\boxed{2}$ $\boxed{3}$ (Fig. 12). The triangular positions of the origin are

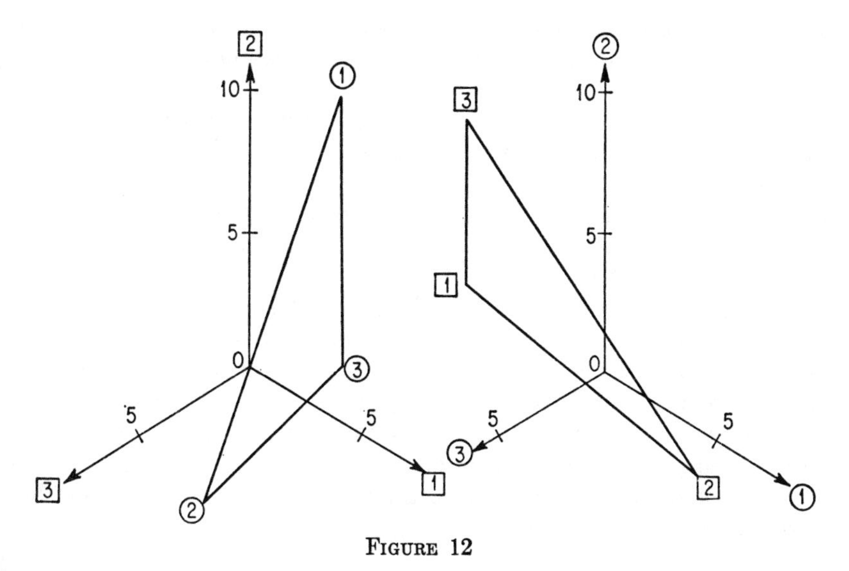

FIGURE 12

$(\frac{1}{3}, \frac{2}{3}, 0)$ and $(\frac{1}{5}, \frac{3}{5}, \frac{1}{5})$. The left-hand figure shows that $\boxed{1}$ dominates a combination of $\boxed{2}$ and $\boxed{3}$, but not strictly. Therefore some solution excludes $\boxed{1}$ and is $(\frac{1}{3}, \frac{2}{3}, 0)$, $(0, \frac{2}{3}, \frac{1}{3})$. The value of the game is 1.

If the matrix is $n \times 3$, the rows may be represented as n points in a triaxial system. The resulting configuration will be useful in picking out which triples of columns to test for a 3×3 solution.

6. MAPPING METHOD FOR SOLVING GAMES WITH CONSTRAINTS

In solving for an optimal strategy of a game we assume that each pure strategy may be played with any probability between zero and one. Aside from the influence of the payoff matrix, we have no preference of strategies or any desire to avoid certain strategies, whether pure or mixed. Now in some situations we may wish to assure that a certain pure strategy is played at least a certain per cent of the time, or that we play a given pure strategy no more frequently than some other given strategy. Such a restriction might arise from the ease or difficulty of playing certain strategies or from the dependence of strategies. If we restrict the choice of strategies in any way, we refer to the game as a *game with constraints*.

We have previously shown that X^*, Y^* is a solution of the game defined by the matrix $A = (a_{ij})$ if and only if

$$\max_X X'AY^* = \min_Y X^{*\prime}AY = X^{*\prime}AY^*.$$

If the game has constraints then the preceding conditions are also necessary and sufficient to define a solution, where X and Y are now the constrained sets of strategies.

The solution X^*, Y^* of an arbitrary game with constraints has the property that it simultaneously maximizes $X'AY^*$ with respect to X and minimizes $X^{*\prime}AY$ with respect to Y. If we consider the set of strategies X and the set of strategies Y, then X^* is some point of X such that the maximum of $X'AY^*$ is assumed at X^*, and Y^* is some point of Y such that the minimum of $X^{*\prime}AY$ is assumed at Y^*. It is this property that enables us to find graphically the solution of a game with constraints.

The method of solution consists of partitioning the sets of strategies X and Y by means of the payoff function. For each partition of X we shall associate some subset Y_1 of Y. Similarly, for each partition of Y we shall associate some subset X_1 of X. X_1, Y_1 will be a solution if X_1 is associated with Y_1 and Y_1 is associated with X_1.

The method can best be described by means of an example. Let the payoff matrix be

$$\begin{array}{c} \\ ① \\ ② \\ ③ \end{array} \begin{array}{c} \begin{array}{ccc} \boxed{1} & \boxed{2} & \boxed{3} \end{array} \\ \begin{bmatrix} 3 & 39 & 30 \\ 33 & 9 & 0 \\ 28 & 4 & 25 \end{bmatrix} \end{array}.$$

We wish to find the solutions of this game such that Blue has the following constraints:

$$\tfrac{1}{10} \leq x_1 \leq \tfrac{4}{5},$$
$$\tfrac{1}{20} \leq x_2 \leq \tfrac{1}{2},$$
$$x_3 \geq \tfrac{6}{5}(1 - \tfrac{20}{9}x_1).$$

This implies that ① must be used at least ten per cent of the time but no more than eighty per cent of the time; ② must be used at least five per cent of the time but no more than fifty per cent of the time; the frequency of using strategy ③ depends on ①, namely $x_3 \geq \frac{6}{5}(1 - \frac{20}{9}x_1)$. Of course, $x_1 + x_2 + x_3 = 1$, always.

Let us also assume that Red has the following constraints:

$$\tfrac{1}{10} \leq y_2 \leq 2y_1, \qquad y_3 \geq \tfrac{1}{6}.$$

The constraints of the first player imply that he picks only those mixed strategies lying within or on the pentagon $ABCDE$ in Fig. 13. The

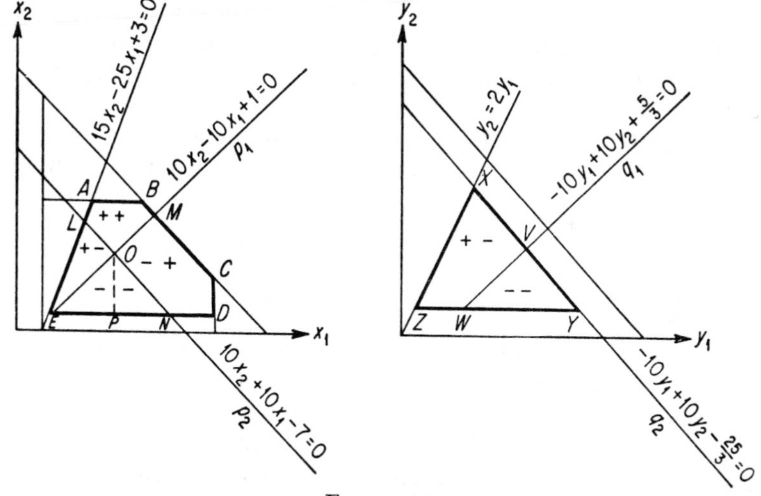

FIGURE 13

second player picks only those mixed strategies lying within or on the triangle XYZ. The boundaries of the pentagon and triangle are determined by the constraints. Now x_1, x_2, $1 - x_1 - x_2$ and y_1, y_2, $1 - y_1 - y_2$ are the probabilities of playing strategies ①, ②, ③ and [1], [2], [3], respectively. The expected payoff is

$$\phi = (10x_2 - 10x_1 + 1)y_1 + (10x_2 + 10x_1 - 7)y_2 + \tfrac{1}{3}(5x_1 - 25x_2 + 25)$$
$$= (-10y_1 + 10y_2 + \tfrac{5}{3})x_1 + (10y_1 + 10y_2 - \tfrac{25}{3})x_2 + (y_1 - 7y_2 + \tfrac{25}{3}).$$

Let us rewrite the expectation ϕ as follows:

$$\begin{aligned}\phi &= p_1y_1 + p_2y_2 + p_3 \\ &= q_1x_1 + q_2x_2 + q_3,\end{aligned}$$

where $p_1 = 10x_2 - 10x_1 + 1$, $q_1 = -10y_1 + 10y_2 + \frac{5}{3}$, etc.

Now $p_1 = 0$, $p_2 = 0$ divide the pentagon $ABCDE$ into four regions: $OLABM$, wherein $p_1 > 0$, $p_2 > 0$; $OMCDN$, wherein $p_1 < 0$, $p_2 > 0$; OEN, wherein $p_1 < 0$, $p_2 < 0$; OEL, wherein $p_1 > 0$, $p_2 < 0$. Similarly, $q_1 = 0$, $q_2 = 0$, divides the triangle XYZ into two regions: $WZXV$, wherein $q_1 > 0$, $q_2 < 0$; VWY, wherein $q_1 < 0$, $q_2 < 0$.

For each point in the region $OLABM$ the expectation ϕ, as a function of Red's strategies, assumes the minimum at point Z in the triangle XYZ. However, for the point Z, the expectation ϕ, as a function of Blue's strategies, assumes its maximum at point D on the pentagon $ABCDE$, and outside the region $OLABM$. Therefore, no point inside region $OLABM$ can provide a solution of the game.

Consider the point O, defined by $p_1 = 0$ and $p_2 = 0$, in the pentagon. For this point ϕ assumes its minimum at every point in the triangle XYZ. In particular, the minimum is assumed at V. The point V is characterized by $q_1 = q_2 = 0$. For the point V, the maximum value of ϕ is assumed at every point of the pentagon and hence at O. Therefore the points O and V are solutions of the game with constraints. These points are the intersection of $p_1 = 0$ with $p_2 = 0$ and the intersection of $q_1 = 0$ with $q_2 = 0$. O is given by

$$x_1 = \tfrac{4}{10}, \quad x_2 = \tfrac{3}{10}, \quad x_3 = \tfrac{3}{10}.$$

V is given by

$$y_1 = \tfrac{3}{6}, \quad y_2 = \tfrac{2}{6}, \quad y_3 = \tfrac{1}{6}.$$

There are additional solutions and they can also be found by the same procedure. Consider the points in region OEN having the property that $p_1 = p_2 \leq 0$. This determines the line $x_1 = \frac{2}{5}$, $x_2 \leq \frac{3}{10}$. For each point on this line the minimum value of ϕ is assumed at each point of XY. In particular, the minimum is assumed at V. But if Red uses strategy V, then ϕ will assume its maximum at every point in the pentagon, and in particular on the line $x_1 = \frac{2}{5}$, $x_2 \leq \frac{3}{10}$. Therefore this line is a solution of the game. It can be verified, by this procedure, that there are no other solutions.

We have found that the solution of this game is:

Blue: $\quad x_1 = \frac{2}{5}, \quad \frac{1}{20} \leq x_2 \leq \frac{3}{10}, \quad x_3 = 1 - x_1 - x_2.$

Red: $\quad y_1 = \frac{1}{2}, \quad y_2 = \frac{1}{3}, \quad y_3 = \frac{1}{6}.$

The value of the game is 6.5.

7. MAPPING METHOD FOR SOLVING GAMES

Of course this mapping or fixed-point method can be used to solve games without constraints. For a 3 × 3 game the strategy space is a right triangle for each player. For example, let the game be defined by the following payoff matrix:

$$A = \begin{bmatrix} -3 & 2 & 0 \\ 0 & 1 & 2 \\ 1 & 2 & 1 \end{bmatrix}.$$

The expectation, ϕ, as a function of a mixed strategy of each player is

$$\begin{aligned}\phi &= (-3x_1 - 2x_2)y_1 + (x_1 - 2x_2 + 1)y_2 + (-x_1 + x_2 + 1) \\ &= (-3y_1 + y_2 - 1)x_1 + (-2y_1 - 2y_2 + 1)x_2 + (y_2 + 1) \\ &= p_1y_1 + p_2y_2 + p_3 \\ &= q_1x_1 + q_2x_2 + q_3,\end{aligned}$$

where

$$\begin{aligned} p_1 &= -3x_1 - 2x_2, & q_1 &= -3y_1 + y_2 - 1, \\ p_2 &= x_1 - 2x_2 + 1, & q_2 &= -2y_1 - 2y_2 + 1, \\ p_3 &= -x_1 + x_2 + 1, & q_3 &= y_2 + 1. \end{aligned}$$

In the Fig. 14, the lines $p_1 = 0$ and $p_2 = 0$ divide the strategy space XWZ of Blue into two regions XYT and $YTWZ$. Similarly, the lines

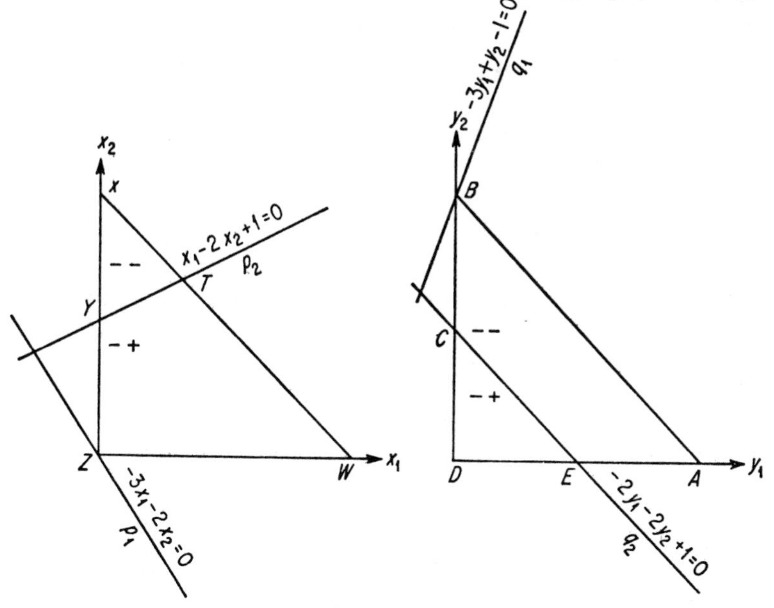

FIGURE 14

$q_1 = 0$, $q_2 = 0$ divide Red's strategy space ABD into two regions $ABCE$ and CDE. The regions are characterized as follows:

$$\begin{aligned} XYT&: \quad p_1 < 0, p_2 < 0 \\ YTWZ&: \quad p_1 < 0, p_2 > 0 \\ ABCE&: \quad q_1 < 0, q_2 < 0 \\ CDE&: \quad q_1 < 0, q_2 > 0. \end{aligned}$$

Consider the point Z for which $p_1 = 0$ and $p_2 = 1$. For this point the minimum of ϕ is assumed at $y_2 = 0$ or the line AD. In particular the minimum is assumed at every point of the line EA. Now each point on line EA has $q_1 < 0$, $q_2 \leq 0$, for which the maximum value of ϕ is assumed at $x_1 = 0$, $x_2 = 0$, or at the point Z. Therefore the point Z is a solution for Blue and the line EA is a solution for Red.

In terms of the strategies, the solutions are:

for Blue: $\qquad x_1 = 0, \qquad x_2 = 0, \qquad x_3 = 1$

for Red: $\qquad \frac{1}{2} \leq y_1 \leq 1, \qquad y_2 = 0, \qquad y_3 = 1 - y_1 - y_2.$

The value of the game is 1.

8. SOLUTION OF RECONNAISSANCE GAME BY MAPPING METHOD

In section 15 of Chapter 3 we showed how the problem of reconnaissance can be considered as a game of strategy. We considered a particular reconnaissance game which yielded the payoff matrix

$$\begin{bmatrix} 46 & 22 & 39 & 27 \\ 10 & 34 & 39 & 27 \\ 55 & 31 & 48 & 24 \\ 19 & 43 & 12 & 36 \end{bmatrix}.$$

We listed the optimal strategies without showing how they were computed. Using the mapping or fixed-point method, we shall now compute the optimal strategies.

If Blue uses a mixed strategy

$$X' = (x_1, x_2, x_3, 1 - x_1 - x_2 - x_3)$$

and Red uses a mixed strategy

$$Y' = (y_1, y_2, y_3, 1 - y_1 - y_2 - y_3),$$

then the expectation of Blue for the above matrix may be written as

$$\begin{aligned} \phi &= p_1 y_1 + p_2 y_2 + p_3 y_3 + p_4 \\ &= q_1 x_1 + q_2 x_2 + q_3 x_3 + q_4, \end{aligned}$$

where

$$\begin{aligned} p_1 &= 36x_1 + 48x_3 - 17 & q_1 &= 36y_1 - 12y_2 + 36y_3 - 9 \\ p_2 &= -12x_1 + 7 & q_2 &= 36y_3 - 9 \\ p_3 &= 36x_1 + 36x_2 + 48x_3 - 24 & q_3 &= 48y_1 + 48y_3 - 12 \\ p_4 &= -9x_1 - 9x_2 - 12x_3 + 36 & q_4 &= -17y_1 + 7y_2 - 24y_3 + 36. \end{aligned}$$

Introduce the following notation:

if $p_1 > 0$, then write

$$p_1 = +,$$

if $q_1 < 0$, then write

$$q_1 = -,$$

if $p_3 < p_2 < p_1 < 0$, then write

$$p_1 = -, \quad p_2 = --, \quad p_3 = ---,$$

if $p_3 > p_2 > p_1 > 0$, then write

$$p_1 = +, \quad p_2 = ++, \quad p_3 = +++.$$

For any point in the strategy space of Blue for which $p_1 = +$, $p_2 = +$, $p_3 = +$, the minimum value of ϕ is assumed at the point $y_1 = y_2 = y_3 = 0$. For the latter point in Red's strategy space, we have

$$q_1 = -9, \quad q_2 = -9, \quad q_3 = -12,$$

or

$$q_1 = -, \quad q_2 = -, \quad q_3 = -.$$

For these values of q_1, the maximum value of ϕ is assumed at $x_1 = x_2 = x_3 = 0$ which yields

$$p_1 = -, \quad p_2 = +, \quad p_3 = -.$$

Therefore no point in Blue's strategy space for which $p_1 > 0$, $p_2 > 0$, $p_3 > 0$ can provide a fixed point or a solution of the game. In a similar way we map each of the possible regions in the space of strategies of Blue. Table 2 summarizes the mapping of the entire space of Blue's strategies.

Using the table, we note that the region defined by $p_1 > 0$, $p_2 > 0$, $p_3 = 0$ maps back into itself. It follows that the optimal strategies for Blue are given by the convex set $X(A)$ defined by

$$36x_1 + 48x_3 - 17 \geq 0,$$

$$-12x_1 + 7 \geq 0,$$

$$36x_1 + 36x_2 + 48x_3 - 24 = 0,$$

$$x_1 \geq 0, \quad x_2 \geq 0, \quad x_3 \geq 0, \quad x_4 = 1 - x_1 - x_2 - x_3.$$

From the table, it is seen that Red has a unique optimal strategy given by

$$y_1 = y_2 = 0, \quad y_3 = \tfrac{1}{4}, \quad y_4 = \tfrac{3}{4}.$$

Table 2. SOLUTION OF RECONNAISANCE GAMES BY MAPPING METHOD

(1) If Blue's strategy is such that			(2) Using (1), expectation, ϕ, assumes its minimum at			(3) Using (2), value of coefficients q_1 at minimum			(4) Using (3), expectation, ϕ, assumes its maximum at			(5) Using (4), value of coefficients p_1 at maximum		
p_1	p_2	p_3	y_1	y_2	y_3	q_1	q_2	q_3	x_1	x_2	x_3	p_1	p_2	p_3
+	+	+	0	0	0	−9	−9	−12	0	0	0	−	+	−
+	+	−	0	0	1	27	27	36	0	0	1	+	+	+
−	−	..	0	0	1	27	27	36	0	0	1	+	+	+
0	0	−	0	0	1	27	27	36	0	0	1	+	+	+
+	−	+	0	1	0	−21	−9	−12	0	0	0	−	+	−
−	..	−	0	1	0	−21	−9	−12	0	0	0	−	+	−
0	−	0	0	1	0	−21	−9	−12	0	0	0	−	+	−
−	+	+	1	0	0	27	−9	36	0	0	1	+	+	+
..	−	−	1	0	0	27	−9	36	0	0	1	+	+	+
−	0	0	1	0	0	27	−9	36	0	0	1	+	+	+
+	..	..	0	α	$1-\alpha$	$27-48\alpha$	$27-36\alpha$	$36-48\alpha$	0				+	
−	..	..	0	α	$1-\alpha$	$27-48\alpha$	$27-36\alpha$	$36-48\alpha$	0				+	
0	..	..	0	α	$1-\alpha$	$27-48\alpha$	$27-36\alpha$	$36-48\alpha$	0				+	
..	+	..	α	0	$1-\alpha$	27	$27-36\alpha$	36	0	0	1	+	+	+
..	−	..	α	0	$1-\alpha$	27	$27-36\alpha$	36	0	0	1	+	+	+
..	0	..	α	0	$1-\alpha$	27	$27-36\alpha$	36	0	0	1	+	+	+

..	..	+	α	$1-\alpha$	0	$48\alpha-21$	-9	$48\alpha-12$	0				+	
..	..	−	α	$1-\alpha$	0	$48\alpha-21$	-9	$48\alpha-12$	0				+	
..	..	0	α	$1-\alpha$	0	$48\alpha-21$	-9	$48\alpha-12$	0				+	
−	−	−	α	β	$1-\alpha-\beta$	$27-48\beta$	$27-36\alpha-36\beta$	$36-48\beta$	0				+	
0	+	+	α	0	0	$36\alpha-9$	-9	$48\alpha-12$		0		p_1	>	p_3
+	0	+	0	α	0	$-12\alpha-9$	-9	-12	0	0	0	−	+	−
+	+	0	0	0	α	$36\alpha-9$	$36\alpha-9$	$48\alpha-12$						
						−	−	− if $\alpha<\frac{1}{4}$	0	0	0	−	+	−
						+	+	+ if $\alpha>\frac{1}{4}$	0	0	1	+	+	+
						0	0	0 if $\alpha=\frac{1}{4}$	a	b	c	+	+	0

Blue's basic solutions are the vertices of the convex set $X(A)$. To determine the vertices we have

$$x_3 = \frac{2 - 3x_1 - 3x_2}{4} \geq 0 \quad \text{or} \quad x_1 + x_2 \leq \tfrac{2}{3}.$$

From the inequalities, we obtain

$$x_2 \leq \tfrac{7}{36}, \qquad x_1 \leq \tfrac{7}{12}.$$

These three inequalities and the equality, together with $x_1 \geq 0$, $x_2 \geq 0$, determine a convex set having five vertices whose coordinates are

$$(0, \tfrac{7}{36}, \tfrac{17}{48}, \tfrac{65}{144}), \quad (\tfrac{17}{36}, \tfrac{7}{36}, 0, \tfrac{1}{3}), \quad (\tfrac{7}{12}, \tfrac{1}{12}, 0, \tfrac{1}{3}),$$
$$(\tfrac{7}{12}, 0, \tfrac{1}{16}, \tfrac{17}{48}), \quad (0, 0, \tfrac{1}{2}, \tfrac{1}{2}).$$

These five vertices are the basic solutions for Blue.

6 GAMES WITH INFINITE NUMBER OF STRATEGIES

1. INTRODUCTION

In a finite game, each player selects a strategy from a finite or discrete set of strategies. The number of such strategies may be large, as in chess, but finite. A natural generalization is to consider games in which a player chooses a strategy from an infinite set of strategies. Such a game is called a *continuous* or infinite game. The name derives from a closed interval's being called a continuum. There is no loss of generality if we assume that the strategies are represented by points on the closed interval [0, 1]. For, if S is the set of strategies, then by relabeling the elements of S, we can get a game in which the selection of a strategy is made from the closed interval [0, 1].

There are several reasons for developing a theory of continuous games. Many military and economic problems, when viewed as games, involve an infinite number of strategies. For example, a military budget can be thought of as being divisible in an infinite number of ways between offense and defense. In economics a commodity may have an infinite number of price possibilities. Further, computations using a continuous variable are sometimes easier than those using a variable which takes on a finite, but large, number of values.

2. DESCRIPTION OF CONTINUOUS GAMES

In its simplest form, a continuous game may be described as follows: Blue chooses a strategy x, where $0 \leq x \leq 1$, and simultaneously Red chooses a strategy y, where $0 \leq y \leq 1$. The choices, x and y, determine a play of the game, whose outcome is measured by a payoff $M(x, y)$ to

Blue. Since the game is assumed to be zero-sum, the payoff to Red is $-M(x, y)$.

Let us assume that the payoff function $M(x, y)$ has a minimum with respect to y for each x and a maximum with respect to x for each y. Suppose Blue chooses some strategy x; then, depending on Red's choice of his strategy, Blue will receive a payoff of at least

$$\min_y M(x, y) = f(x). \tag{6.1}$$

Now Blue can choose x to make $f(x)$ as large as possible. Hence there is a way for Blue to play or there exists a strategy for Blue, so that he receives at least

$$\max_x \min_y M(x, y).$$

Similarly, there exists a strategy for Red such that Blue receives at most

$$\min_y \max_x M(x, y).$$

It is easily shown that these two quantities satisfy the inequality

$$\max_x \min_y M(x, y) \leq \min_y \max_x M(x, y). \tag{6.2}$$

If in (6.2) the equality holds, or that

$$\max_x \min_y M(x, y) = \min_y \max_x M(x, y) = M(x_0, y_0), \tag{6.3}$$

then $M(x, y)$ has a saddle-point at x_0, y_0. Since the strategies x_0 and y_0 satisfy the conditions

$$\begin{aligned} M(x_0, y) &\geq M(x_0, y_0) \qquad \text{for all } y, \\ M(x, y_0) &\leq M(x_0, y_0) \qquad \text{for all } x, \end{aligned} \tag{6.4}$$

we call x_0 an optimal strategy for Blue and y_0 an optimal strategy for Red. Thus, if a continuous game has a saddle-point, then the saddle-point yields a pair of optimal strategies, one for each player.

3. MIXED STRATEGY—DISTRIBUTION FUNCTION

Suppose the game is such that

$$\max_x \min_y M(x, y) < \min_y \max_x M(x, y); \tag{6.5}$$

then the game does not have a saddle-point. It will be recalled that if a finite game does not have a saddle-point it is necessary to introduce mixed strategies, which are probability distribution functions over the finite set of strategies. In order to solve infinite games without saddle-points it will be necessary to introduce probability distribution functions over the infinite set of strategies. However, it is no longer possible to think of a mixed strategy as a rule which ascribes a probability to each number in the closed interval [0, 1]. We need a different definition of probability for infinite number of strategies. Of course, any definition we formulate for the infinite game must also apply to the finite game.

A mixed strategy in an infinite game is a random process for choosing a number x from the interval [0, 1]. We can also think of a mixed strategy as a random process for choosing a number no larger than x from the interval [0, 1]. If ξ is a random variable such that $0 \leq \xi \leq 1$, then a mixed strategy is a function F such that, for all x in [0, 1],

$$F(x) = \text{pr}\ \{\xi \leq x\}. \tag{6.6}$$

That is, $F(x)$ is the probability that the number chosen by the random process F is at most x. For mathematical convenience, we shall modify the definition slightly for the case where $x = 0$. We shall define $F(0) = 0$. Thus $F(0)$ is the probability that the number chosen will be actually less than zero, not at most zero. Suppose $0 < x_1 < x_2 \leq 1$; then we see that

$$F(x_2) - F(x_1) = \text{pr}\ \{x_1 < \xi \leq x_2\}$$

and

$$F(x_2) - F(0) = \text{pr}\ \{0 \leq \xi \leq x_2\}.$$

The function F is called a *cumulative distribution function*, or simply a *distribution function.*

From our definition of F in (6.6) as a probability function it follows that F has the following properties:

(i) $F(x) \geq 0$ for all $0 \leq x \leq 1$.
(ii) $F(0) = 0$, $F(1) = 1$.
(iii) F is a nondecreasing function of x.

$$F(x_1) \leq F(x_2) \quad \text{whenever } x_1 \leq x_2.$$

(iv) F is right-hand continuous in the open interval (0, 1), or

$$\lim_{\epsilon \to 0} F(x + \epsilon) = F(x).$$

The last result follows from the fact that

$$F(x + \epsilon) - F(x) = \text{pr}\ \{x < \xi \leq x + \epsilon\}.$$

Now for any given ξ it is possible to find an $\epsilon > 0$ which is sufficiently small so that the above inequality cannot hold. Hence the probability that ξ satisfies this inequality approaches zero as ϵ approaches zero.

Any function F which satisfies the above four conditions is a distribution function. It is obvious that for a finite set of strategies we can get a mixed strategy from the distribution function over the finite set of strategies.

We can construct new distribution functions from given ones as follows:

Let $F_1, F_2, \ldots, F_n$ be n distribution functions; then by forming any convex linear combination of these functions,

$$F(x) = \alpha_1 F_1(x) + \ldots + \alpha_n F_n(x),$$

where $\alpha_i \geq 0$ and $\Sigma \alpha_i = 1$, we get another distribution function.

Although a distribution function must be right-hand continuous, it may be discontinuous. A common class of distribution functions which are discontinuous are the *step-functions*. Define the following function:

$$\begin{aligned} I_a(x) &= 0 \quad \text{if } x < a, \\ &= 1 \quad \text{if } x \geq a. \end{aligned}$$

It is obvious that if $0 \leq a \leq 1$, then $I_a(x)$ is a distribution function. It has a discontinuity at $x = a$, with a jump of 1 at a. We call $I_a(x)$ a step-function with one step at a.

The distribution function

$$F(x) = \alpha_1 I_{x_1}(x) + \alpha_2 I_{x_2}(x) + \ldots + \alpha_n I_{x_n}(x),$$

where $\alpha_i \geq 0$, $\Sigma \alpha_i = 1$, $0 \leq x_i \leq x_{i+1} \leq 1$, is a step-function with n steps.

To assist us in computing a distribution function $F(x)$, we make use of the following definition of choosing a point at random on a line: If a point is chosen at random on a line interval of length l, the probability that it is in a given subinterval of length λ is λ/l.

4. EXPECTATION—STIELTJES INTEGRAL

Having defined a mixed strategy as a probability distribution function, we now need to define expectation. Suppose Blue receives a payoff $P(x)$ if he chooses strategy x. Suppose further that Blue chooses his strategy x by means of a random device which picks x according to the probability distribution function $F(x)$. We wish to determine the expectation of $P(x)$ with respect to the probability distribution function $F(x)$.

It is clear that we cannot say this expectation is simply the sum of all products $P(x)F(x)$. On the other hand we may approximate the expectation by dividing the interval [0, 1] into a large number of subintervals, computing the expectation over each subinterval, and adding these expectations for all the subintervals. Suppose we divide the interval [0, 1] by the points

$$0 = x_0 < x_1 < x_2 < \ldots < x_n = 1.$$

Let
$$\Delta = \max_{1 \leq i \leq n} (x_i - x_{i-1})$$

and let
$$x_{i-1} \leq Z_i \leq x_i \qquad \text{for } 1 \leq i \leq n.$$

Then we can approximate the expectation of $P(x)$ by

$$\sum_{i=1}^{n} P(Z_i)[F(x_i) - F(x_{i-1})].$$

If

$$\lim_{\substack{n\to\infty \\ \Delta\to 0}} \sum_{i=1}^{n} P(Z_i)[F(x_i) - F(x_{i-1})] \tag{6.7}$$

exists and is independent of the choice of Z_i, then this limit is called the *Stieltjes integral* of P with respect to F from 0 to 1 and is denoted by

$$\int_0^1 P(x)\, dF(x). \tag{6.8}$$

We say that if $P(x)$ is Blue's payoff corresponding to a strategy choice x, then if he chooses x according to the distribution function F, his expectation will be

$$E(P(x)) = \int_0^1 P(x)\, dF(x).$$

The Stieltjes integral does not always exist. In particular, if P and F have a common point of discontinuity, then the integral does not exist. For example, let

$$P(x) = F(x) = 0 \qquad \text{for } 0 \leq x < \tfrac{1}{2},$$
$$P(x) = F(x) = 1 \qquad \text{for } \tfrac{1}{2} \leq x \leq 1.$$

Here exactly one of the differences $F(x_i) - F(x_{i-1})$ is different from 0, call it $F(x_k) - F(x_{k-1})$, which is equal to 1. Then $P(Z_k) = 0$ or 1 depending on the choice Z_k. Thus the limit (6.7) is not independent of the choice of the Z_i's and hence (6.8) does not exist.

5. STIELTJES INTEGRAL FOR CONTINUOUS FUNCTION

We shall now show that if P is continuous in the interval $[0, 1]$, then

$$\int_0^1 P(x)\, dF(x)$$

exists.

Suppose the interval $[0, 1]$ is divided into n intervals by the points $0 = x_0 < x_1 < x_2 < \ldots < x_n = 1$. Let the interval $[x_{k-1}, x_k]$ be denoted by H. Define the following quantities:

$$M_k = \max_H P(x),$$

$$m_k = \min_H P(x),$$

$$S_n = \sum_{i=1}^{n} M_i[F(x_i) - F(x_{i-1})],$$

$$s_n = \sum_{i=1}^{n} m_i[F(x_i) - F(x_{i-1})].$$

It is evident that

$$s_n \leq S_n \qquad \text{for all } n.$$

The sequence $\{S_n\}$ decreases with increasing n. For, let the interval $[0, 1]$ be divided into n parts. To obtain a division of the interval $[0, 1]$

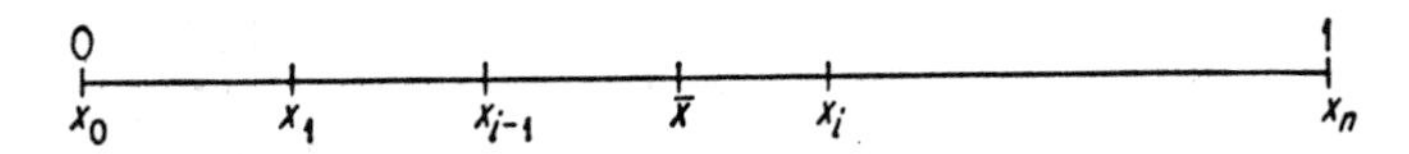

into $n + 1$ parts let us introduce one more division point, $\bar{x}$, say it falls between x_i and x_{i-1}, i.e., $x_{i-1} < \bar{x} < x_i$.

Define

$$\overline{M}_i = \max_{\bar{x} < x \leq x_i} P(x) \leq M_i,$$

$$\overline{\overline{M}}_i = \max_{x_{i-1} < x \leq \bar{x}} P(x) \leq M_i.$$

Then

$$S_{n+1} = \sum_{i=1}^{n+1} M_i[F(x_i) - F(x_{i-1})],$$

$$\begin{aligned} S_n - S_{n+1} &= M_i[F(x_i) - F(x_{i-1})] - \overline{\overline{M}}_i[F(\bar{x}) - F(x_{i-1})] \\ &\qquad - \overline{M}_i[F(x_i) - F(\bar{x})] \\ &\geq M_i[F(x_i) - F(x_{i-1}) - F(\bar{x}) + F(x_{i-1}) - F(x_i) + F(\bar{x})] \\ &= M_i \cdot 0 = 0. \end{aligned}$$

Thus

$$S_n \geq S_{n+1}.$$

The sequence $\{s_n\}$ increases with increasing n. For, let us make a similar division of the interval $[0, 1]$ as in the previous case and define

$$\overline{m}_i = \min_{\bar{x} < x \leq x_i} P(x) \geq m_i,$$

$$\overline{\overline{m}}_i = \min_{x_{i-1} < x \leq \bar{x}} P(x) \geq m_i.$$

Then

$$s_{n+1} = \sum_{i=1}^{n+1} m_i[F(x_i) - F(x_{i-1})],$$

$$\begin{aligned} s_{n+1} - s_n &= \overline{m}_i[F(x_i) - F(\bar{x})] + \overline{\overline{m}}_i[F(\bar{x}) - F(x_{i-1})] \\ &\qquad - m_i[F(x_i) - F(x_{i-1})] \\ &\geq m_i[F(x_i) - F(\bar{x}) + F(\bar{x}) - F(x_{i-1}) - F(x_i) + F(x_{i-1})] = 0. \end{aligned}$$

Therefore

$$s_{n+1} \geq s_n.$$

Combining the inequalities on s_n and S_n we have

$$s_n \leq s_{n+1} \leq S_{n+1} \leq S_n \leq S_1$$

or

$$s_n \leq S_1 \qquad \text{for all } n.$$

Therefore the increasing sequence $\{s_n\}$ is bounded. Hence it has a limit. Let this limit be defined by

$$s = \lim_{\substack{n \to \infty \\ \Delta \to 0}} s_n$$

where Δ is the length of the largest of the n intervals.

We also have

$$S_{n+1} \geq s_{n+1} \geq s_n > s_1.$$

The decreasing sequence $\{S_n\}$ has a greatest lower bound

$$S = \lim_{\substack{n \to \infty \\ \Delta \to 0}} S_n.$$

Since P is continuous over the closed interval $[0, 1]$, it is uniformly continuous. Hence corresponding to any positive number ϵ, there exists a positive number δ such that if $|z_1 - z_2| < \delta$ then

$$|P(z_1) - P(z_2)| < \epsilon.$$

This means that if $\Delta < \delta$ then $|M_k - m_k| < \epsilon$ for $k = 1, 2, \ldots, n$. Thus

$$0 < S_n - s_n = \sum_{i=1}^{n} (M_i - m_i)[F(x_i) - F(x_{i-1})]$$

$$\leq \epsilon \sum_{i=1}^{n} [F(x_i) - F(x_{i-1})] = \epsilon.$$

We have then

$$0 \leq S_n - s_n = (S_n - S) + (S - s) + (s - s_n) \leq \epsilon.$$

Each of the three terms in the parentheses is ≥ 0. Since ϵ can be made arbitrarily small, it follows that each of the three terms must approach 0 in the limit, and

$$S = s.$$

Now for any choice of the z_i's we have

$$s_n \leq \Sigma\, P(z_i)[F(x_i) - F(x_{i-1})] \leq S_n.$$

Since

$$\lim_{n \to \infty} s_n = \lim_{n \to \infty} S_n$$

it follows that

$$\lim_{\substack{n \to \infty \\ \Delta \to 0}} \Sigma\, P(z_i)[F(x_i) - F(x_{i-1})] = S$$

exists and is independent of the choice of the z_i's.

6. STIELTJES INTEGRAL AND RIEMANN INTEGRAL

In some cases a Stieltjes integral may be evaluated like the ordinary, or Riemann, integral. In particular, if

$$\int_0^1 P(x)\, dF(x)$$

exists and if F has a derivative F' at every point in $[0, 1]$ then we have

$$\int_0^1 P(x)\, dF(x) = \int_0^1 P(x)F'(x)\, dx.$$

Hence, in this case, we can evaluate the Stieltjes integral, if it exists, by evaluating a well-known Riemann integral, if it exists.

Let the interval $[0, 1]$ be divided into n parts by the points $0 = x_0 < x_1 < x_2 < \ldots < x_n = 1$. Let Δ be the length of the largest interval. Since F has a derivative everywhere, there exist y_i where $x_{i-1} \leq y_i \leq x_i$ $(i = 1, 2, \ldots, n)$ such that

$$F(x_i) - F(x_{i-1}) = F'(y_i)(x_i - x_{i-1}). \tag{6.9}$$

Now if the Riemann integral $\int_0^1 P(x)F'(x)\, dx$ exists it is defined by

$$\lim_{\substack{n\to\infty \\ \Delta\to 0}} \sum_{i=1}^{n} P(y_i)F'(y_i)(x_i - x_{i-1}) = \int_0^1 P(x)F'(x)\, dx \tag{6.10}$$

independent of the choice of the y_i's. Substituting (6.9) into (6.10) we get

$$\lim_{\substack{n\to\infty \\ \Delta\to 0}} \sum_{i=1}^{n} P(y_i)[F(x_i) - F(x_{i-1})] = \int_0^1 P(x)F'(x)\, dx \tag{6.11}$$

where the left side of (6.11) is independent of the choice of the y_i's. But the left side of (6.11) is the definition of the Stieltjes integral. Hence we have

$$\int_0^1 P(x)\, dF(x) = \int_0^1 P(x)F'(x)\, dx.$$

7. STIELTJES INTEGRAL WITH RESPECT TO A STEP-FUNCTION

If $F(x)$ is a step-function, or

$$F(x) = I_\alpha(x),$$

and if $P(x)$ is continuous at α, then

$$\int_0^1 P(x)\, dI_\alpha(x) = P(\alpha).$$

The result above follows from the continuity of $P(x)$ at $x = \alpha$. Let the points of division of $[0, 1]$ be $0 = x_0 < x_1 < x_2 < \ldots < x_n = 1$. Let $x_{j-1} < \alpha \leq x_j$. Then

$$F(x_j) - F(x_{j-1}) = 1 \qquad \text{for } i = j,$$
$$F(x_i) - F(x_{i-1}) = 0 \qquad \text{for } i \neq j.$$

Therefore

$$\sum_{i=1}^{n} P(z_i)[F(x_i) - F(x_{i-1})] = P(z_j)$$

where $x_{j-1} \leq z_j \leq x_j$. But from the continuity of $P(z)$ at $z = \alpha$, we have that

$$\lim_{\substack{n\to\infty \\ \Delta\to 0}} P(z_j) = P(\alpha).$$

Therefore

$$\lim_{\substack{n\to\infty \\ \Delta\to 0}} \sum_{i=1}^{n} P(z_i)[F(x_i) - F(x_{i-1})] = P(\alpha) \tag{6.12}$$

independent of the choice of the z_i's. The left side of (6.12) is the definition of a Stieltjes integral, hence

$$\int_0^1 P(x)\, dI_\alpha(x) = P(\alpha).$$

Using a similar argument we can show that if

$$F(x) = a_1 I_{\alpha_1}(x) + a_2 I_{\alpha_2}(x) + \ldots + a_n I_{\alpha_n}(x)$$

where $a_i \geq 0$, $\sum_{i=1}^{n} a_i = 1$, $0 \leq \alpha_i \leq 1$, and if $P(x)$ is continuous at the n values, $\alpha_1, \alpha_2, \ldots, \alpha_n$, then

$$\int_0^1 P(x)\, dF(x) = a_1 P(\alpha_1) + a_2 P(\alpha_2) + \ldots + a_n P(\alpha_n).$$

8. SOME PROPERTIES OF STIELTJES INTEGRAL

If the integrals involved exist, it can be readily shown that they have the following properties:

(i) $\int_0^1 P(x)\, dF(x) = \int_0^a P(x)\, dF(x) + \int_a^1 P(x)\, dF(x).$

(ii) $\int_0^1 [P(x) + Q(x)]\, dF(x) = \int_0^1 P(x)\, dF(x) + \int_0^1 Q(x)\, dF(x).$

(iii) $\int_0^1 P(x)\, d[kF(x)] = k \int_0^1 P(x)\, dF(x).$

(iv) $\int_0^1 P(x)\, d[F(x) + G(x)] = \int_0^1 P(x)\, dF(x) + \int_0^1 P(x)\, dG(x).$

(v) $\int_0^1 dF(x) = F(1) - F(0) = 1.$

(vi) $\int_0^1 P(x)\, dF(x) = P(1)F(1) - P(0)F(0) - \int_0^1 F(x)\, dP(x).$

(vii) $\left|\int_0^1 P(x)\, dF(x)\right| \le \int_0^1 |P(x)|\, dF(x).$

(viii) If $P(x) \le Q(x)$ in $[0, 1]$ then

$$\int_0^1 P(x)\, dF(x) \le \int_0^1 Q(x)\, dF(x).$$

7 SOLUTION OF INFINITE GAMES

1. OPTIMAL MIXED STRATEGY

Suppose that the payoff function is $M(x, y)$, and suppose Blue chooses his strategy x from $[0, 1]$ using the distribution function $F(x)$. Then for any strategy y chosen by Red, the expectation of Blue, if it exists, is given by

$$E(F, y) = \int_0^1 M(x, y)\, dF(x).$$

Now suppose that Red chooses y by means of the distribution function $G(y)$; then the expectation of Blue, if it exists, will be

$$E(F, G) = \int_0^1 \int_0^1 M(x, y)\, dF(x)\, dG(y).$$

$E(F, G)$ is Blue's expectation if he plays a strategy determined by the distribution function $F(x)$ and Red plays a strategy determined by the distribution function $G(y)$.

Suppose the following two expressions exist:

$$\max_F \min_G E(F, G) = v_1,$$

$$\min_G \max_F E(F, G) = v_2.$$

Then there exists a distribution function F^* such that Blue can receive (in expectation) at least v_1. There also exists a distribution function G^* such that Blue gets at most v_2.

In general, $v_1 \leq v_2$. However, if $v_1 = v_2$, or if

$$\max_F \min_G E(F, G) = \min_G \max_F E(F, G) = v, \tag{7.1}$$

then we call v the *value* of the game to Blue. Further, there exists an F^* such that Blue receives at least v regardless of Red's mixed strategy, i.e.,

$$\min_G E(F^*, G) = v.$$

Therefore

$$E(F^*, G) \geq v \qquad \text{for all } G. \tag{7.2}$$

Similarly, there exists a G^* such that

$$E(F, G^*) \leq v \qquad \text{for all } F. \tag{7.3}$$

Thus F^*, G^* are called *optimal mixed strategies* for Blue and Red, respectively. The pair F^*, G^* is also called a *solution* of the game. We also have

$$E(F^*, G^*) = v. \tag{7.4}$$

2. EXISTENCE OF OPTIMAL STRATEGIES

From the minimax theorem on finite games, it follows that every game with a finite number of strategies has a solution. However, every infinite game does not have a solution. There are examples of infinite games which do not have solutions. However, if the payoff function $M(x, y)$ is continuous in each of the two strategic variables, then it can be proven that the game always has a solution. Thus if $M(x, y)$ is continuous in x and y, there exists a pair of distribution functions F^*, G^*, one for each player such that

$$\max_F \int_0^1 \int_0^1 M(x, y)\, dF(x)\, dG^*(y) = \min_G \int \int M(x, y)\, dF^*(x)\, dG(y).$$

We can also state that F^*, G^* is a solution and v is the value of the game, if and only if

$$\max_x \int_0^1 M(x, y)\, dG^*(y) = \min_y \int_0^1 M(x, y)\, dF^*(x) = v. \tag{7.5}$$

To prove (7.5), let us first show that if $P(x)$ is continuous, then

$$\max_F \int_0^1 P(x)\, dF(x) = \max_x P(x).$$

Since $P(x)$ is continuous on the interval $[0, 1]$, it has a maximum. Let the maximum occur at a, or

$$\max_x P(x) = P(a).$$

Hence

$$P(a) \geq P(x) \qquad \text{for all } x \text{ in } [0, 1].$$

For any distribution function F, it follows that

$$\int_0^1 P(a)\, dF(x) \geq \int_0^1 P(x)\, dF(x).$$

Hence for any F

$$P(a) \geq \int_0^1 P(x)\, dF(x).$$

Let us compute the least upper bound of the numbers

$$\int P(x)\, dF(x).$$

That is, we wish

$$\sup_F \int P(x)\, dF(x).$$

Now

$$\sup_F \int P(x)\, dF(x) \geq \int_0^1 P(x)\, dI_a(x) = P(a).$$

But, for any F

$$P(a) \geq \int_0^1 P(x)\, dF(x).$$

In particular,

$$P(a) \geq \sup_F \int_0^1 P(x)\, dF(x).$$

Hence

$$\sup_F \int_0^1 P(x)\, dF(x) = P(a) = \int_0^1 P(x)\, dI_a(x).$$

Thus the least upper bound, with respect to F, of the numbers $\int P(x)\, dF(x)$ is actually assumed at $F = I_a$. Therefore

$$\max_F \int_0^1 P(x)\, dF(x) = P(a) = \max_x P(x).$$

Similarly, we can show

$$\min_G \int_0^1 P(y)\, dG(y) = \min_y P(y).$$

Now (7.5) follows from the fact that

$$\begin{aligned}\max_F \int_0^1 \int_0^1 M(x, y)\, dF(x)\, dG^*(y) &= \int_0^1 \left[\max_F \int_0^1 M(x, y)\, dF(x)\right] dG^*(y)\\ &= \int_0^1 \max_x M(x, y)\, dG^*(y)\\ &= \max_x \int_0^1 M(x, y)\, dG^*(y)\end{aligned}$$

and $$\min_G \int_0^1 \int_0^1 M(x, y)\, dG(y)\, dF^*(x) = \min_y \int_0^1 M(x, y)\, dF^*(x).$$

3. PROPERTIES OF OPTIMAL STRATEGIES

Let $H^*(x)$ be Blue's expectation if he uses a pure strategy x and Red uses an optimal strategy $G^*(y)$, i.e.,

$$H^*(x) = \int_0^1 M(x, y)\, dG^*(y).$$

Similarly, let $K^*(y)$ be Blue's expectation if Red uses a pure strategy y and Blue uses an optimal strategy F^*, i.e.,

$$K^*(y) = \int_0^1 M(x, y)\, dF^*(x).$$

The following properties are readily proven:

(i) $H^*(x) \leq v \leq K^*(y)$ for all x and y in $[0, 1]$. Each pure strategy when used against an opponent's optimal mixed strategy cannot yield a higher expectation than an optimal strategy.

(ii) $\max_x H^*(x) = \min_y K^*(y) = v$. Each player has at least one pure strategy which, if used against an opponent's optimal strategy, yields the value of the game.

(iii) If $H^*(x_0) < v$ then $\text{pr}\,\{\xi = x_0\} = 0$. A player's optimal mixed strategy contains no strategy which yields less than the value of the game when that strategy is used against the opponent's optimal strategy. If $K^*(y_0) > v$ then $\text{pr}\,\{\xi = y_0\} = 0$.

(iv) If $\text{pr}\,\{\xi = x_0\} > 0$ then $H^*(x_0) = v$. Each pure strategy in the mixed strategy must yield the value of the game when used against the opponent's optimal mixed strategy.

Example. Suppose the payoff is given by

$$M(x, y) = (x - y)^2.$$

Let us verify that the optimal strategy for Blue is to play $x = 0$ and $x = 1$ at random with equal probability and the optimal strategy for Red is to play $y = \frac{1}{2}$. We wish to verify that

$$F^*(x) = \tfrac{1}{2}I_0(x) + \tfrac{1}{2}I_1(x), \qquad G^*(y) = I_{1/2}(y).$$

Evaluating, we have

$$\max_x \int_0^1 M(x, y)\, dG^*(y) = \max_x M(x, \tfrac{1}{2})$$
$$= \max_x (x - \tfrac{1}{2})^2 = \tfrac{1}{4}.$$

$$\min_y \int_0^1 M(x, y)\, dF^*(x) = \min_y [\tfrac{1}{2}M(0, y) + \tfrac{1}{2}M(1, y)]$$
$$= \min_y [\tfrac{1}{2}y^2 + \tfrac{1}{2}(1 - y)^2] = \tfrac{1}{4}.$$

Since

$$\max_x \int_0^1 M(x, y)\, dG^*(y) = \min_y \int_0^1 M(x, y)\, dF^*(x) = \tfrac{1}{4},$$

it follows that F^*, G^* as defined above are optimal strategies and $v = \frac{1}{4}$.

We also have, for this game, that

$$H^*(x) = \int_0^1 M(x, y)\, dG^*(y) = (x - \tfrac{1}{2})^2$$

$$K^*(y) = \int_0^1 M(x, y)\, dF^*(x) = \tfrac{1}{2}[y^2 + (1 - y)^2].$$

Further

$$\max_x H^*(x) = H^*(0) = H^*(1) = \tfrac{1}{4}.$$

$$\min_y K^*(y) = K^*(\tfrac{1}{2}) = \tfrac{1}{4}.$$

$$H^*(x) < \tfrac{1}{4} \qquad \text{for all } x \neq 0, 1$$

$$k^*(y) > \tfrac{1}{4} \qquad \text{for all } y \neq \tfrac{1}{2}.$$

If ξ is the random variable selected by the optimal mixture, note that

for Blue: $\qquad \text{pr}\,\{\xi \neq 0, 1\} = 0,$

for Red: $\qquad \text{pr}\,\{\xi \neq \tfrac{1}{2}\} = 0.$

Example. We now give an example of a game which has a continuous solution. Suppose the payoff is given by

$$M(x, y) = |y - x|\,(1 - |y - x|).$$

Let us verify that a solution of this game is $F^*(x) = x$ and $G^*(y) = y$. We have

$$H^*(x) = \int_0^1 M(x, y)\, dG^*(y) = \int_0^1 |y - x|\,(1 - |y - x|)\, dy$$

$$= \int_{y=0}^{x} (x - x^2 + 2xy - y - y^2)\, dy$$

$$+ \int_{y=x}^{y=1} (y - y^2 + 2xy - x - x^2)\, dy = \tfrac{1}{6}.$$

$$K^*(y) = \int_0^1 M(x, y)\, dF^*(x)$$

$$= \int_0^1 |y - x|\,(1 - |y - x|)\, dx = \tfrac{1}{6}.$$

Since

$$H^*(x) \equiv \tfrac{1}{6} = K^*(y) \qquad \text{for all } x \text{ and } y$$

it follows that $F^*(x) = x$, $G^*(y) = y$ are optimal strategies and $v = \tfrac{1}{6}$.

4. DELAYED FIRING

An example of a game with a discontinuous payoff which has a solution is the operational problem of scheduling the firing of a missile requiring an exposure time. Suppose that Blue plans to fire one missile before T hours have expired. However, in order to fire the missile, Blue must expose it for e hours, where $e < T$, during which time the missile is vulnerable to attack by Red. Let us assume that Red, not knowing Blue's decision, has only one opportunity to attack. When is the optimal time for Blue to expose his missile? When is the optimal time for Red to attack?

A strategy for Blue is a choice of time X for him to begin exposing the missile for a time e, where $0 \leq X \leq T - e$. Since Red does not know Blue's choice, a strategy for Red is a choice of time Y for Red to attack Blue, where $0 \leq Y \leq T$. Blue will fire his missile at time $X + e$ if he has not been attacked by Red during the time of missile exposure. That is, if Red attacks either before or after missile exposure, Blue will be able to fire his missile.

Let the payoff to Blue be 1 if he fires the missile (i.e., Blue's missile was not attacked during exposure) and 0 otherwise. Then in terms of the strategies of the players the payoff is described by the following discontinuous payoff function:

$$M(X, Y) = \begin{cases} 1 & \text{if } Y \leq X \leq T - e \text{ or } X + e \leq Y \leq T \\ 0 & \text{otherwise.} \end{cases}$$

Let $t = e/T$, $x = X/T$, and $y = Y/T$; then Blue's strategies now range over the interval $[0, 1 - t]$ and Red's strategies now range over the interval $[0, 1]$. The payoff function becomes

$$M(x, y) = \begin{cases} 1 & \text{if } y \leq x \leq 1 - t \text{ or } x + t \leq y \leq 1 \\ 0 & \text{otherwise.} \end{cases} \tag{7.6}$$

Let us assume that $t \leq \frac{1}{2}$; then it is readily shown that

$$0 = \sup_x \inf_y M(x, y) < \inf_y \sup_x M(x, y) = 1.$$

Hence if a solution exists it must be in terms of mixed strategies. Let $F(x)$ and $G(y)$ be mixed strategies of Blue and Red, respectively. The expected payoff to Blue is

$$\begin{aligned} E(F, G) &= \int_0^{1-t} \int_0^1 M(x, y)\, dG(y)\, dF(x) \\ &= \int_0^{1-t} \int_0^x dG(y)\, dF(x) + \int_0^{1-t} \int_{x+t}^1 dG(y)\, dF(x) \qquad (7.7) \\ &= \int_0^{1-t} [G(x) + 1 - G(x + t)]\, dF(x). \end{aligned}$$

Using (7.7), we have

$$\begin{aligned} \inf_G \sup_F E(F, G) &= \inf_G \sup_{0 \leq x \leq 1-t} [G(x) + 1 - G(x + t)] \\ &= 1 - \sup_G \inf_{0 \leq x \leq 1-t} [G(x + t) - G(x)]. \end{aligned}$$

Now let

$$\alpha(G) = \inf_{0 \leq x \leq 1-t} [G(x + t) - G(x)].$$

Then, for all x, we have

$$\alpha(G) \leq G(x+t) - G(x)$$

or

$$G(x+t) \geq G(x) + \alpha(G) \qquad \text{for all } 0 \leq x \leq 1-t.$$

Setting $x = 0, t, 2t, \ldots, (n-1)t$, we obtain

$$\begin{aligned} G(t) &\geq \alpha(G) \\ G(2t) &\geq G(t) + \alpha(G) \geq 2\alpha(G) \\ G(3t) &\geq 3\alpha(G) \\ &\cdot \\ &\cdot \\ &\cdot \\ G(nt) &\geq n\alpha(G), \end{aligned}$$

where $nt \leq 1 < (n+1)t$, i.e., $n = [1/t]$ or the largest integer contained in $1/t$.

Therefore, we have

$$n\alpha(G) \leq G(nt) \leq G(1) = 1$$

and

$$\alpha(G) \leq \frac{1}{n} \qquad \text{for all } G.$$

It follows that

$$\sup_G \inf_{0 \leq x \leq 1-t} [G(x+t) - G(x)] \leq \frac{1}{n}. \tag{7.8}$$

We shall show that the equality holds in (7.8). We do this by exhibiting a function G which yields the sup. Suppose that $1/t$ is an integer, or $nt = 1$. If we take $G(y) = y$, we have

$$\inf_{0 \leq x \leq 1-t} [G(x+t) - G(x)] = t = \frac{1}{n}.$$

Therefore

$$\sup_G \inf_x [G(x+t) - G(x)] = \frac{1}{n}.$$

Now suppose that $1/t$ is not an integer, and let $[1/t] = n$. Consider the distribution function

$$G(y) = \frac{1}{n} \sum_{j=1}^{n} I_{j/(n+1)}(y).$$

Then

$$G(x+t) - G(x) = \frac{1}{n} \sum_{j=1}^{n} [I_{j/(n+1)}(x+t) - I_{j/(n+1)}(x)].$$

For this function $G(y)$ we have

$$\inf_{0 \leq x \leq 1-t} [G(x+t) - G(x)] = \frac{1}{n}.$$

Hence, we have

$$\sup_G \inf_{0 \leq x \leq 1-t} [G(x+t) - G(x)] = \frac{1}{n} = \frac{1}{[1/t]}.$$

$$\inf_G \sup_F E(F, G) = 1 - \frac{1}{[1/t]}\cdot$$

gument we can show that

$$\sup_F \inf_G E(F, G) = 1 - \frac{1}{[1/t]}\cdot$$

Therefore the game has a value given by

$$v = 1 - \frac{1}{[1/t]}\cdot$$

An optimal strategy for Blue is

$$F^*(x) = \frac{1}{n} \sum_{j=0}^{n-1} I_{jt}(x).$$

To verify that F* is optimal, we consider two cases. First, suppose $t = 1/n$, or $1/t$ is an integer; then

$$\begin{aligned}
E(F^*, G) &= \int_0^{1-t} [G(x) + 1 - G(x + t)]\, dF^*(x) \\
&= 1 - \int_0^{1-t} [G(x + t) - G(x)]\, dF^*(x) \\
&= 1 - \frac{1}{n} \sum_{j=0}^{n-1} \int_0^{1-t} [G(x + t) - G(x)]\, dI_{jt}(x) \\
&= 1 - \frac{1}{n} \sum_{j=0}^{n-1} [G(jt + t) - G(jt)] \\
&= 1 - \frac{1}{n} [G(1) - G(0)] = 1 - \frac{1}{n} = 1 - \frac{1}{[1/t]}\cdot
\end{aligned}$$

Suppose $\dfrac{1}{n+1} < t < \dfrac{1}{n}$; then for all G,

$$\begin{aligned}
E(F^*, G) &= \int_0^{1-t} [G(x) + 1 - G(x + t)]\, dF^*(x) \\
&= 1 - \frac{1}{n} \sum_{j=0}^{n-1} [G(jt + t) - G(jt)] \\
&= 1 - \frac{1}{n} [G(nt) - G(0)] \geq 1 - \frac{1}{n} = 1 - \frac{1}{[1/t]}\cdot
\end{aligned}$$

In terms of the original parameters, the solution of the game is as follows:

(i) The value of the game $= 1 - \dfrac{1}{[T/e]}\cdot$

(ii) Blue's optimal strategy is given by

$$F^*(X) = \frac{1}{[T/e]} \sum_{j=0}^{[T/e]-1} I_{je/T}(X).$$

(iii) Red's optimal strategy is given by

$$G^*(Y) = Y, \quad \text{if } T/e \text{ is an integer,}$$

$$G^*(Y) = \frac{1}{[T/e]} \sum_{j=1}^{T/e} I_{j/([T/e]+1)}(Y) \quad \text{if } T/e \text{ is not an integer.}$$

5. EXAMPLE OF GAME WITHOUT SOLUTION

We shall now give an example of a game with a discontinuous payoff function which does not have a solution, nor a value.

Consider the infinite game in which each player chooses a number in the closed interval [0, 1]. Let x and y represent the choices by Blue and Red, respectively. Define the payoff to Blue as follows:

$$M(x, y) = \begin{cases} 0 & \text{for } x = y; \\ -1 & \text{for } x = 1, y < 1; \text{ and } x < y < 1; \\ +1 & \text{for } y = 1, x < 1; \text{ and } y < x < 1. \end{cases}$$

The above payoff may be represented as in Fig. 15 on the unit square.

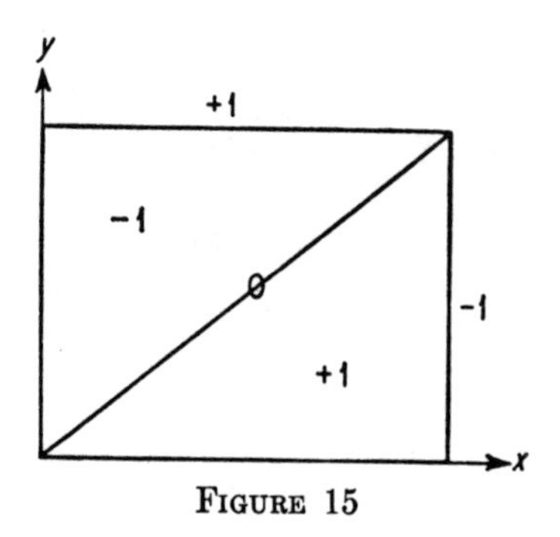

FIGURE 15

Let us compute Blue's expectation $E(F, y)$ if Blue uses the mixed strategy F and Red uses the pure strategy y. We have

(i) if $0 < y < 1$,

$$\begin{aligned} E(F, y) &= \int_{x=0}^{1} M\, dF = -1 \int_{0}^{y-0} dF + 1 \int_{y+0}^{1-0} dF - 1 \int_{1-0}^{1} dF \\ &= -F(y-0) + F(1-0) - F(y) - F(1) + F(1-0) \\ &= -1 + 2F(1-0) - [F(y-0) + F(y)]; \end{aligned}$$

(ii) if $y = 0$,

$$E(F, y) = +1 \int_{0+0}^{1-0} dF - 1 \int_{1-0}^{1} dF$$
$$= 2F(1-0) - F(1) = 2F(1-0) - 1;$$

(iii) if $y = 1$,

$$E(F, y) = +1 \int_{0}^{1-0} dF = F(1-0).$$

Since $F(y) \geq F(y-0)$, we have from the above that if $0 < y < 1$,

$$E(F, y) \leq -1 + 2F(1-0) - 2F(y-0)$$
$$= -1 + 2[F(1-0) - F(y-0)].$$

Therefore, for any F and any ϵ, $(0 < \epsilon < \frac{1}{2})$, there exists a y_0 such that

$$E(F, y_0) \leq -1 + \epsilon.$$

In other words, for any F, there exists a G, namely $G = I_{y_0}(y)$ such that

$$E(F, G) \leq -1 + \epsilon.$$

Therefore

$$\inf_G E(F, G) \leq -1 + \epsilon$$

and

$$\sup_F \inf_G E(F, G) \leq -1 + \epsilon.$$

Similarly, we can show that for any G and any ϵ, such that $(0 < \epsilon < \frac{1}{2})$ there exists an F such that

$$E(F, G) \geq +1 - \epsilon.$$

Therefore

$$\inf_G \sup_F E(F, G) \geq 1 - \epsilon.$$

Thus we have shown that for this game, we have

$$\inf_G \sup_F E(F, G) > \sup_F \inf_G E(F, G).$$

8 GAMES WITH CONVEX PAYOFF FUNCTIONS

1. CONVEX PAYOFF FUNCTIONS

For many military and economic games the continuous payoff function is also convex in one variable. For example, the outcome of many types of attack-defense games is naturally described by a convex payoff function. In such cases it is possible to describe the form of the optimal strategies.

A function f is called *convex* in the interval $[0, 1]$ if, for every λ for which $0 \leq \lambda \leq 1$, and for every pair of strategies y_1, y_2, we have

$$f[\lambda y_1 + (1 - \lambda)y_2] \leq \lambda f(y_1) + (1 - \lambda)f(y_2).$$

If the equality never holds for $\lambda \neq 0$, $\lambda \neq 1$, we call f *strictly convex*. Geometrically, if a function is convex then between any two points of the graph of the function, the graph never lies above the segment connecting the two points. The function is strictly convex if the graph of the function always lies below the line segment. If a convex function is not strictly convex, its graph consists in part of straight line segments.

Suppose f is a function of n variables; then f is convex if for every pair of distinct points $(y_1, y_2, \ldots, y_n)$ and $(\overline{y}_1, \overline{y}_2, \ldots, \overline{y}_n)$ on the interval, we have

$$f[\lambda y_1 + (1 - \lambda)\overline{y}_1, \ldots, \lambda y_n + (1 - \lambda)\overline{y}_n] \leq \lambda f(y_1, \ldots, y_n) + (1 - \lambda)f(\overline{y}_1, \ldots, \overline{y}_n).$$

We call f strictly convex if the equality never holds for $\lambda \neq 0$, $\lambda \neq 1$.

Suppose f is twice differentiable; then f is also convex in y if

$$\frac{d^2f}{dy^2} \geq 0 \qquad \text{for all } y.$$

The function f is strictly convex if

$$\frac{d^2f}{dy^2} > 0 \qquad \text{for all } y.$$

Let us assume that $M(x, y)$ is strictly convex in y for each x and that $M(x, y)$ is twice differentiable (with respect to y). Then for any distribution function $F(x)$, it follows that the function

$$K(y) = \int_0^1 M(x, y)\, dF(x)$$

is also strictly convex. For, if $\dfrac{\partial^2 M}{\partial y^2} > 0$ for all y, then

$$\frac{\partial^2 K(y)}{\partial y^2} = \int \frac{\partial^2 M(x, y)}{\partial y^2}\, dF(x) > 0.$$

2. OPTIMAL PURE STRATEGY FOR RED

Let us assume that $M(x, y)$ is strictly convex in y for each x and is continuous in both variables. Suppose the game has a solution, which will be verified. Let $F^*(x)$ be an optimal strategy for Blue. Then Blue may announce the choice $F^*(x)$ to Red, who will pick some G^* such that

$$\begin{aligned}\iint M(x, y)\, dF^*(x)\, dG^*(y) &= \min_G \iint M(x, y)\, dF^*(x)\, dG(y)\\ &= \min_y \int M(x, y)\, dF^*(x)\\ &= \min_y K^*(y).\end{aligned}$$

Now $K^*(y)$ is strictly convex in y. Hence $K^*(y)$ assumes its minimum at one point. Therefore an optimal strategy for Red is a pure strategy, the y which minimizes $K^*(y)$.

3. VALUE OF GAME IS $\min_y \max_x M(x, y)$

Continuing with the assumption that $M(x, y)$ is continuous and strictly convex in y, we can obtain the value of the game. For any game, its value is given by

$$v = \min_G \max_x \int M(x, y)\, dG(y).$$

We have shown that Red's optimal strategy is a pure strategy. Hence we need consider only those G's which are one-step distribution functions. Therefore

$$v = \min_y \max_x \int_0^1 M(x, y)\, dI_y(y) = \min_y \max_x M(x, y).$$

4. RED'S OPTIMAL PURE STRATEGY

An optimal strategy y^* for Red must be such that

$$v = \min_y \max_x M(x, y) = \max_x M(x, y^*).$$

Thus y^* has the property that it minimizes $\max_x M(x, y)$. From the strict convexity of $M(x, y)$ it follows that y^* is unique.

5. BLUE'S OPTIMAL STRATEGIES

Blue's optimal strategy may be pure or mixed, depending on the location of Red's optimal strategy. If Red's optimal strategy is an end point of the interval [0, 1], i.e., if $y^* = 0$ or 1, then Blue has a pure strategy. If Red's optimal strategy y^* is such that $0 < y^* < 1$, then an optimal strategy for Blue is to randomize over two values of x.

Let Red's optimal strategy be an end point of the interval [0, 1], e.g., $y^* = 0$. Then

$$\max_x M(x, 0) = v.$$

Hence

$$M(x, 0) \leq v \qquad \text{for all } x.$$

Let X_0 be the set of points x_0 such that

$$M(x_0, 0) = v \qquad \text{for all } x_0 \text{ in } X_0.$$

Then the remaining set of points X_1 of the interval [0, 1] have the property that

$$M(x_1, 0) < v \qquad \text{for all } x_1 \text{ in } X_1.$$

Now an optimal strategy for Blue consists of mixing some strategies x_0 satisfying the condition $M(x_0, 0) = v$. We shall show that there exists an optimal pure strategy for Blue, i.e., there exists some x_0 such that $M(x_0, y) \geq v$ for all y. This is equivalent to showing that there exists some $x_0 \in X_0$ such that

$$v = \min_y M(x_0, y) = M(x_0, 0)$$

or that $M(x_0, y)$ is a nondecreasing function at $y = 0$.

Suppose that for every $x_0 \in X_0$, the function $M(x_0, y)$ is decreasing at $y = 0$, or

$$\frac{\partial M(x_0, 0)}{\partial y} = M'(x_0, 0) < 0.$$

Then for sufficiently small $\epsilon > 0$, and every x_0 in X_0,

$$M(x_0, y) < v \qquad \text{for } 0 < y < \epsilon.$$

Since $M(x_1, y)$ is continuous in y, it follows that for sufficiently small $\epsilon > 0$, and every x_1 in X_1,

$$M(x_1, y) < v \qquad \text{for } 0 < y < \epsilon.$$

Therefore, for every x in $[0, 1]$,

$$M(x, y_1) < v \qquad 0 < y_1 < \epsilon.$$

Hence

$$\max_x M(x, y_1) < v \qquad 0 < y_1 < \epsilon.$$

This yields the contradiction

$$v = \min_y \max_x M(x, y) \leq \max_x M(x, y_1) < v.$$

It follows that there exists some $x_0 \in X_0$ such that the function $M(x_0, y)$ is nondecreasing at $y = 0$, or

$$M'(x_0, 0) \geq 0.$$

Since $M(x_0, 0) = v$ and $M(x_0, y)$ is convex in y, it follows that

$$v = \min_y M(x_0, y) \quad \text{or} \quad M(x_0, y) \geq v \qquad \text{for all } y.$$

We have shown that if $y^* = 0$, then Blue has an optimal pure strategy x^* satisfying two conditions:

$$M(x^*, 0) = v, \qquad M'(x^*, 0) \geq 0.$$

In a similar manner we can show that if $y^* = 1$, then Blue has an optimal pure strategy x^* satisfying two conditions:

$$M(x^*, 1) = v, \qquad M'(x^*, 1) \leq 0.$$

Now suppose that $0 < y^* < 1$. Then

$$M(x, y^*) \leq v \qquad \text{for all } x.$$

Let X_0 be the set for which

$$M(x_0, y^*) = v \qquad \text{for all } x_0 \in X_0,$$

and let X_1 be the set for which

$$M(x_1, y^*) < v \qquad \text{for all } x_1 \in X_1.$$

Suppose that every x_0 in X_0 were such that

$$M'(x_0, y^*) < 0;$$

then we would be led to the same contradiction as for the case $y^* = 0$. Hence there exists some x_0^* such that

$$M(x_0^*, y^*) = v, \qquad M'(x_0^*, y^*) \geq 0.$$

In a similar manner we can show that there exists some x_{00}^* such that

$$M(x_{00}^*, y^*) = v, \qquad M'(x_{00}^*, y^*) \leq 0.$$

Now consider the function

$$f(t) = tM'(x_0^*, y^*) + (1 - t)M'(x_{00}^*, y^*).$$

We note that

$$f(0) = M'(x_{00}^*, y^*) \leq 0,$$
$$f(1) = M'(x_0^*, y^*) \leq 0.$$

Since f is a continuous function of t, there exists an α, where $0 \leq \alpha \leq 1$, such that

$$\alpha M'(x_0^*, y^*) + (1 - \alpha)M'(x_{00}^*, y^*) = 0.$$

Having determined x_0^*, x_{00}^{**}, and α we shall now show that an optimal strategy for Blue is

$$F^*(x) = \alpha I_{x_0^*}(x) + (1 - \alpha)I_{x_{00}^*}(x).$$

We have that

$$K^*(y) = \int M(x, y)\, dF^*(x) = \alpha M(x_0^*, y) + (1 - \alpha)M(x_{00}^*, y).$$

Since $M(x, y)$ is convex in y, it follows that $K^*(y)$ is convex in y. Further

$$\frac{dK^*(y)}{dy} = \alpha M'(x_0^*, y) + (1 - \alpha)M'(x_{00}^*, y)$$

which vanishes at $y = y^*$. Hence $K^*(y)$ assumes its minimum at y^*, or

$$\min_y K^*(y) = \alpha M(x_0^*, y^*) + (1 - \alpha)M(x_{00}^*, y^*) = v.$$

Therefore $F^*(x)$ is an optimal strategy for Blue.

We have shown that if $0 < y^* < 1$, then an optimal strategy for Blue is the step-function

$$F^*(x) = \alpha I_{x_1}(x) + (1 - \alpha)I_{x_2}(x)$$

where

$$M(x_1, y^*) = M(x_2, y^*) = v,$$
$$M'(x_1, y^*) \geq 0 \geq M'(x_2, y^*),$$
$$\alpha M'(x_1, y^*) + (1 - \alpha)M'(x_2, y^*) = 0.$$

6. CONCAVE PAYOFF FUNCTIONS

The following dual result for concave payoff functions can be proven in the same way as for convex functions:

Suppose that $M(x, y)$ is continuous in both variables and is a strictly concave function of x for each y. Let $\partial M(x, y)/\partial x$ exist for each x and y in the unit interval. The solution of the game is as follows:

(i) $v = \max_x \min_y M(x, y)$.

(ii) Blue has a unique optimal pure strategy x^*.

(iii) (a) If $x^* = 0$, then Red has an optimal pure strategy y^* such that $M(0, y^*) = v$ and $\partial M(0, y^*)/\partial x \leq 0$.

(b) If $x^* = 1$, then Red has an optimal pure strategy y^* such that $M(1, y^*) = v$ and $\partial M(1, y^*)/\partial x \geq 0$.

(c) If $0 < x^* < 1$, then Red has an optimal mixed strategy which is of the form

$$G^*(y) = \alpha I_{y_1}(y) + (1 - \alpha) I_{y_2}(y)$$

where the parameters α, y_1, y_2 satisfy the conditions

$$M(x^*, y_1) = M(x^*, y_2) = v,$$

$$\frac{\partial M(x^*, y_1)}{\partial x} \leq 0 \leq \frac{\partial M(x^*, y_2)}{\partial x},$$

$$\alpha \frac{\partial M(x^*, y_1)}{\partial x} + (1 - \alpha) \frac{\partial M(x^*, y_2)}{\partial x} = 0.$$

7. GENERAL CONVEX PAYOFF

In the previous discussion we assumed that the payoff was strictly convex or strictly concave. Most of the results are still valid if the strictness assumption is removed. However, it is no longer true that the optimal strategies are unique.

Although we have assumed that each player's strategy space is one-dimensional, similar arguments would prove analogous results for n-dimensional strategy space.

Example. Suppose the payoff function is

$$M(x, y) = (x - y)^2.$$

Since

$$\frac{\partial^2 M(x, y)}{\partial^2 y^2} = 2,$$

it follows that $M(x, y)$ is convex in y for each x. Hence the value of the game is

$$v = \min_y \max_x (x - y)^2 = \min_y \max [y^2, (1 - y)^2] = \tfrac{1}{4}.$$

Red's optimal strategy is defined by that pure strategy which minimizes $\max [y^2, (1 - y)^2]$. Therefore $y^* = \frac{1}{2}$. Blue's optimal strategy is a mixture of those pure strategies x_i such that $M(x_i, \frac{1}{2}) = v$, or $(x - \frac{1}{2})^2 = \frac{1}{4}$. This yields $x = 0$ and 1. To obtain the weights α and $1 - \alpha$, we solve the equation

$$\alpha M'(0, \tfrac{1}{2}) + (1 - \alpha) M'(1, \tfrac{1}{2}) = 0.$$

Substituting, we obtain $\alpha = \frac{1}{2}$. Therefore Blue's optimal strategy is

$$F^*(x) = \tfrac{1}{2} I_0(x) + \tfrac{1}{2} I_1(x).$$

8. DEFENSE OF TWO TARGETS AGAINST ATTACK

Suppose that Red, the Defender, has two targets T_1 and T_2 of values k_1 and k_2, respectively, which are to be defended against an attack by Blue. Let us assume Red and Blue are equally strong, i.e., they have the same number of forces, S.

A strategy for Blue is an allocation of x attacking units to T_1, where $0 \leq x \leq S$, and the remainder, $S - x$, to T_2. A strategy for Red is an allocation of y to T_1, where $0 \leq y \leq S$, and $S - y$ to T_2.

	Target T_1	*Target* T_2
Value	k_1	k_2
Offense allocation	x	$S - x$
Defense allocation	y	$S - y$

Let the payoff to Blue be proportional to the number of attacking units that get through to target and the target value. Thus if $x \geq y$, then $S - x \leq S - y$, in which case we assume that $x - y$ units survive to hit target T_1 and none survive to hit target T_2. In this case the payoff to the attack is $k_1(x - y)$. If $x \leq y$ then $S - x \geq S - y$, and $y - x$ units survive to attack target T_2 while none survive to hit target T_1. The payoff can be summarized as follows:

$$M(x, y) = \begin{cases} k_1(x - y) & \text{if} \quad x \geq y \\ k_2(y - x) & \text{if} \quad x \leq y. \end{cases}$$

One may interpret k_1 to be the payoff per attacking unit that penetrates the defenses at T_1.

It is evident that $M(x, y)$ is a convex function of y for each x. It consists of two lines as shown in Fig. 16. Therefore

$$v = \min_y \max_x M(x, y) = \min_y \max \, [k_2 y, k_1(S - y)].$$

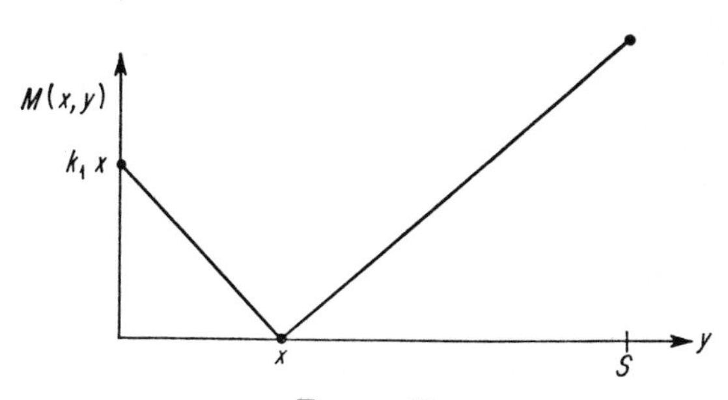

FIGURE 16

Now the function

$$\max\,[k_2y, k_1(S - y)]$$

assumes its minimum at that y for which

$$k_2y = k_1(S - y) \quad \text{or} \quad y^* = \frac{k_1S}{k_1 + k_2}.$$

Therefore Red's optimal strategy is to allocate $\frac{k_1}{k_1 + k_2}S$ defense units to T_1 and the remainder, $\frac{k_2}{k_1 + k_2}S$ defense units, to T_2.

The value of the game is

$$v = \frac{k_1k_2}{k_1 + k_2}S.$$

Solving for Blue's optimal strategy, we set

$$M\left(x, \frac{k_1S}{k_1 + k_2}\right) = v.$$

This equation yields two solutions:

$$x_1 = 0, \qquad x_2 = S.$$

Now

$$\frac{\partial M(0, y^*)}{\partial y} = k_2, \qquad \frac{\partial M(S, y^*)}{\partial y} = -k_1.$$

Hence

$$F^*(x) = \alpha I_0(x) + (1 - \alpha)I_S(x)$$

where

$$\alpha k_2 + (1 - \alpha)(-k_1) = 0 \quad \text{or} \quad \alpha = \frac{k_1}{k_1 + k_2}.$$

We summarize the solution of this game as follows: The Defender splits his forces and may adopt a fixed deployment of such forces—place $k_1/(k_1 + k_2)$ of them at T_1 and $k_2/(k_1 + k_2)$ of them at T_2. The Attacker's optimal strategy is mixed. The Attacker concentrates his forces on either T_1 or T_2 chosen at random. He chooses T_1 with probability $k_2/(k_1 + k_2)$ and he chooses T_2 with probability $k_1/(k_1 + k_2)$.

For example, if target T_2 is three times as valuable as target T_1, or $k_2 = 3k_1$, then T_2 is defended by three-fourths of the defensive force. The optimal strategy for the Attacker is to strike T_1 with all his forces with 0.75 probability.

9. DEFENSE OF MANY TARGETS OF DIFFERENT VALUES

An example of the application of the results on an n-dimensional convex payoff function is a simple air defense problem. Like most battle situations, the combat between air attack and air defense can be viewed as a zero-sum two-person game: The attacker seeks the greatest possible gains in the form of the destruction of targets, and the defender wishes to make these gains as small as possible.

An important decision of the defender in a battle situation is the distribution of his total defense resources among his targets. An important decision of the attacker is the distribution of his total attacking force among those targets. We shall consider this game in a very simplified form in which we assume only a single choice for each player, namely, for the attacker the choice of an allocation of his resources among targets, and for the defender the choice of an allocation of his resources among those targets.

We wish to answer such questions as: Should all the targets be defended? If only some of the targets are to be defended, how shall these be selected? How should the attacker select his targets?

To answer these questions we consider the following game model:

Defense. The defender, Red, has D units of defense to distribute among his n targets, which we label $T_1, T_2, \ldots, T_n$. Let us assume that the n targets have values $k_1, k_2, \ldots, k_n$, respectively, and are ordered as follows:

$$0 < k_1 < k_2 < \ldots < k_n.$$

Attack. The attacker, Blue, has A units of attack to distribute among the n targets. Let us assume the attack is at least as strong as the defense, or $A \geq D$.

Strategy. A strategy for Blue is an allocation of his attacking force A among the n targets. Thus a strategy for Blue is a set of numbers $x_1, x_2, \ldots, x_n$ such that

$$x_i \geq 0 \quad \text{and} \quad \sum_{i=1}^{n} x_i = A.$$

A strategy for Red is a set of numbers $y_1, y_2, \ldots, y_n$ such that

$$y_i \geq 0 \quad \text{and} \quad \sum_{i=1}^{n} y_i = D.$$

Each y_i represents the number of defensive units allocated to target T_i.

Payoff. Let us assume that one unit of defense can check one unit of attack. Further, let us assume that the amount of damage to any target is proportional to the number of attacking units which outnumber the defending units, the coefficient of proportionality depends on the particular target. Finally, let us assume that the payoff is the sum, over the targets, of the damage to each target. Thus the payoff to Blue is

$$M(x, y) = \sum_{i=1}^{n} k_i \max(0, x_i - y_i)$$

where

$$x_i \geq 0, \quad y_i \geq 0, \quad \sum_{i=1}^{n} x_i = A, \quad \text{and} \quad \sum_{i=1}^{n} y_i = D.$$

Solution. It is apparent that $M(x, y)$ is convex in y for each x. It is also convex in x, for each y. Therefore Red, the defender, has a pure strategy which is optimal. The attacker has a mixed strategy which is optimal.

It can be verified that it is optimal for the defender to distribute his defensive force D among the high-valued targets. It can also be verified that it is optimal for the attacker to select one of the high-valued targets at random, subject to a given probability distribution, and then allocate his entire attacking force on that target.

To give a more precise description of the optimal strategies for the two players, we introduce the following notation:

$$\frac{1}{h_s} = \sum_{i=s}^{n} \frac{1}{k_i}, \qquad s = 1, 2, \ldots, n;$$

$$l_s = k_s - h_s\left(n - s + 1 - \frac{D}{A}\right), \qquad s = 1, 2, \ldots, n;$$

$$m = \text{smallest value of } s \text{ such that } l_s \geq 0.$$

In terms of the above definitions, we can verify the following:

The attacker's optimal mixed strategy is:

(a) Never attack the low-valued targets $T_1, T_2, \ldots, T_{m-1}$.

(b) Use the entire attacking force A on a target selected at random, subject to the following probability distribution:

$$\text{pr}\,\{x_i = A\} = \frac{h_m}{k_i}, \qquad m \leq i \leq n.$$

The defender's optimal pure strategy is:

(a) Leave undefended the low-valued targets $T_1, T_2, \ldots, T_{m-1}$.

(b) Defend the high-valued targets $T_m, T_{m+1}, \ldots, T_n$ by placing

$$A\left\{1 - \frac{h_m}{k_i}\left(n - m + 1 - \frac{D}{A}\right)\right\}, \qquad m \leq i \leq n,$$

units at the ith target.

The value of the game to the attacker is

$$v = A(k_m - l_m).$$

It is of interest to note that if $m \leq i \leq n$, then

$$\begin{aligned} k_i \max(0, A - y_i) &= k_i A \frac{h_m}{k_i}\left(n - m + 1 - \frac{D}{A}\right) \\ &= h_m A\left(n - m + 1 - \frac{D}{A}\right) \\ &= A(k_m - l_m) = v. \end{aligned}$$

Thus at each of the defended targets, the attacker gets the value of the game if that target is attacked by the entire attacking force.

If $1 \leq i \leq m - 1$, then

$$k_i \max(0, A - y_i) = k_i A < A(k_m - l_m).$$

Thus, at each of the undefended targets, a concentrated attack yields less than the value of the game. If the defender has allocated his defenses optimally, there is no soft spot in his targets.

Example. Suppose the defender has five targets to defend against an attack. Let the target values be the following:

$$k_1 = \tfrac{1}{12}, \quad k_2 = \tfrac{1}{9}, \quad k_3 = \tfrac{1}{7}, \quad k_4 = \tfrac{1}{5}, \quad k_5 = \tfrac{1}{4}.$$

Suppose that the attacker and the defender have equal forces, or $A = D = S$. We have

$$\frac{1}{h_1} = \sum_{i=1}^{5} \frac{1}{k_i} = 12 + 9 + 7 + 5 + 4 = 37,$$

$$l_1 = k_1 - h_1(5 - 1 + 1 - 1) = \frac{1}{12} - \frac{4}{37} < 0,$$

$$\frac{1}{h_2} = \sum_{i=2}^{5} \frac{1}{k_i} = 9 + 7 + 5 + 4 = 25,$$

$$l_2 = k_2 - h_2(5 - 2 + 1 - 1) = \frac{1}{9} - \frac{3}{25} < 0,$$

$$\frac{1}{h_3} = \sum_{i=3}^{5} \frac{1}{k_i} = 7 + 5 + 4 = 16,$$

$$l_3 = k_3 - h_3(5 - 3 + 1 - 1) = \frac{1}{7} - \frac{2}{16} > 0.$$

Since $l_3 > 0$ and $l_2 < 0$, it follows that $m = 3$. Therefore the optimal allocations are such that the first two targets are not defended, nor are they attacked. Targets 3, 4, and 5 are defended as follows:

$$S\left\{1 - \frac{7}{16}(5 - 3 + 1 - 1)\right\} = \frac{S}{8} \qquad \text{defense units at target } T_3,$$

$$S\left\{1 - \frac{5}{16}(2)\right\} = \frac{3}{8}S \qquad \text{defense units at target } T_4,$$

$$S\left\{1 - \frac{4}{16}(2)\right\} = \frac{S}{2} \qquad \text{defense units at target } T_5.$$

The optimal strategy for the attacker is to select one of the three targets T_3, T_4, or T_5 at random and concentrate his attack on that target. The probabilities associated with these targets are $\frac{7}{16}, \frac{5}{16}, \frac{4}{16}$, respectively.

The value of the game is

$$v = A(k_m - l_m) = S\left(\frac{1}{7} - \frac{1}{7} + \frac{2}{16}\right) = \frac{S}{8}.$$

9 GAMES OF TIMING—DUELS

1. DUEL AS A GAME OF TIMING

The theory of games may be used to analyze a class of problems dealing with the timing of decisions in a competitive environment. In these problems, the actions which the players may take are given in advance, but the timing of the actions is by the strategic decisions of the players. Such games are characterized by the following conflict of interests: each player wishes to delay his decision as long as possible, but he may be penalized for waiting. In a duel, for example, each duelist wishes to hold his fire as long as possible, since his accuracy increases with time. However, if the duelist holds his fire too long, his opponent may win the duel.

Since the duel is a good example of a game of timing, we shall use the duel with bullets as our model of games of timing. Thus, we shall consider an action to be the firing of a bullet. The result of the action is given by an accuracy function representing the probability of hitting the opponent as a function of time of firing. As in all games, we need to describe the information available to the players. If a duelist is informed about his opponent's actions as soon as they take place, we shall call the duel a *noisy duel.* If neither duelist ever learns when or whether his opponent has fired, we shall call the duel a *silent duel.*

2. NOISY DUEL: ONE BULLET EACH DUELIST

Let us first consider the noisy duel in which each duelist has one bullet. Each duelist is informed of his opponent's action, firing his bullet, as soon as it takes place. Further, let us assume that if a duelist fires and misses, the other duelist can obtain a sure hit by waiting until they are together. The duelists, starting at a distance D apart, approach each other with no opportunity for retreat. The accuracies increase steadily as the duelists approach each other and ultimately are certainty, or 1, when the duelists are breast to breast.

A strategy for Blue instructs him when to fire his bullet if his opponent has not already fired; and if the opponent has already fired and missed, then Blue fires when his accuracy is 1. Thus a strategy for Blue is to fire his bullet when the duelists are x units apart, where $0 \leq x \leq D$. Similarly, a strategy for Red is to fire when the duelists are y units apart, where $0 \leq y \leq D$. Let the accuracies of Blue and Red be $P_1(x)$ and $P_2(y)$, respectively. That is, $P_1(x)$ is the probability of Blue's hitting his opponent if he fires when the duelists are x units apart. Assume the accuracies increase as the distance decreases.

Let the payoff be $+1$ to the surviving duelist and 0 to each duelist if both survive or neither survives. The payoff $M(x, y)$, to Blue, is his expectation of survival for his three possible ranges of firing times: firing before Red fires, firing at the same time as Red fires, and firing after Red fires. Thus the payoff is given by

$$M(x, y) = \begin{cases} P_1(x)(1) + [1 - P_1(x)](-1) = 2P_1(x) - 1 & \text{if } x > y \\ P_1(x)[1 - P_2(x)] + P_2(x)[1 - P_1(x)](-1) & \\ \quad = P_1(x) - P_2(x) & \text{if } x = y \\ P_2(y)(-1) + [1 - P_2(y)](1) = 1 - 2P_2(y) & \text{if } y > x. \end{cases}$$

Since $P_1(x)$ and $P_2(y)$ increase with decreasing values of x and y, respectively, it follows that

$$\max_x \min_y M(x, y) = \max_x \min\,[2P_1(x) - 1, P_1(x) - P_2(x), 1 - 2P_2(x)].$$

Now divide the interval $[0, D]$ into three intervals as follows:

Interval	*Consists of those x for which*
A	$P_1(x) + P_2(x) \geq 1$
B	$P_1(x) + P_2(x) = 1$
C	$P_1(x) + P_2(x) \leq 1$

These intervals are not vacuous.

Let

$$\mu(x) = \min\,[2P_1(x) - 1, P_1(x) - P_2(x), 1 - 2P_2(x)].$$

Then

$$\max_x \min_y M(x, y) = \max_x \mu(x) = \max\,[\max_{x \varepsilon A} \mu(x), \max_{x \varepsilon B} \mu(x), \max_{x \varepsilon C} \mu(x)].$$

Now for all x in A, we have

$$P_1(x) + P_2(x) \geq 1,$$

from which it follows that

$$1 - 2P_2(x) \leq P_1(x) - P_2(x) \leq 2P_1(x) - 1.$$

Therefore if $x \,\varepsilon\, A$, then

$$\mu(x) = 1 - 2P_2(x).$$

In interval B, which is a point, we have

$$P_1(x) + P_2(x) = 1.$$

It follows that

$$1 - 2P_2(x) = P_1(x) - P_2(x) = 2P_1(x) - 1.$$

Therefore if $x \in B$, then

$$\mu(x) = P_1(x) - P_2(x).$$

Interval C is defined by those x for which

$$P_1(x) + P_2(x) \leq 1.$$

It follows that

$$2P_1(x) - 1 \leq P_1(x) - P_2(x) \leq 1 - 2P_2(x).$$

Therefore, for all x in C,

$$\mu(x) = 2P_1(x) - 1.$$

Let x^* be defined by

$$P_1(x^*) + P_2(x^*) = 1.$$

It follows that

$$\max_{x \in A} \mu(x) = 1 - 2P_2(x^*),$$

$$\max_{x \in B} \mu(x) = P_1(x^*) - P_2(x^*),$$

$$\max_{x \in C} \mu(x) = 2P_1(x^*) - 1.$$

Therefore, we have

$$\max_x \min_y M(x, y) = P_1(x^*) - P_2(x^*)$$

where x^* satisfies the equation

$$P_1(x^*) + P_2(x^*) = 1.$$

In a similar manner we can show that

$$\min_y \max_x M(x, y) = P_1(y^*) - P_2(y^*)$$

where y^* satisfies the equation

$$P_1(y^*) + P_2(y^*) = 1.$$

We have thus shown that $M(x, y)$ has a saddle-point at x^*, y^*. The optimal strategy for each player is to fire when he is at a distance l from his opponent given by

$$P_1(l) + P_2(l) = 1.$$

The value of the game is $P_1(l) - P_2(l)$.

Summary of solution. The optimal strategy for the duelists is to fire their bullets simultaneously at a distance x_0 which satisfies the equation

$$P_1(x_0) + P_2(x_0) = 1.$$

If Blue uses this strategy, then he is sure of receiving at least $P_1(x_0) - P_2(x_0)$. If Red uses this strategy, he will lose at most $P_1(x_0) - P_2(x_0)$.

Example. Suppose that Blue's accuracy is given by $P_1(x) = 1 - x$ and Red's accuracy is given by $P_2(y) = 1 - y^2$. Then each duelist should fire his bullet at distance x determined by

$$x + x^2 = 1$$

or $x = 0.62$. The value of this duel is $x^2 - x = -0.24$ to Blue and $+0.24$ to Red.

If the two duelists have the same accuracies, then they should fire when their accuracies are 0.5. The value of this duel game is zero.

3. NOISY DUEL: ONE BULLET EACH DUELIST, WITHOUT SADDLE-POINT

In the previous section we discussed a noisy duel which had a saddle-point for its solution, i.e., each player had an optimal pure strategy. The existence of the saddle-point depends largely on the assumption that the duelists have equal worths. In particular we assumed a payoff of one unit to the surviving duelist, whether he be Blue or Red. If the payoff to the surviving duelist depends on which duelist survives, then the resulting duel does not have a saddle-point. A duel between a bomber and a fighter, where the bomber is worth more than the fighter, is an example of a duel with unequal worths.

Let us assume, as before, that the two duelists approach each other. Their accuracies are $P_1(x)$ and $P_2(y)$, respectively. We assume that the duel is noisy, i.e., if a duelist misses, his opponent is certain of a hit. Let us assume the following payoff to Blue:

α, if Blue alone survives,
β, if Red alone survives,
γ, if neither Blue nor Red survives,
0, if both Blue and Red survive.

It is reasonable to assume that $\alpha > \beta$.

Let a strategy for each duelist be a time to fire if his opponent has not fired, and to fire at the time when his accuracy is 1 if his opponent has fired and missed. If x and y are strategies of Blue and Red, respectively, then the payoff to Blue is:

$$M(x, y) = \begin{cases} (\alpha - \beta)P_1(x) + \beta & \text{if } x < y, \\ \alpha P_1(x) + \beta P_2(x) + (\gamma - \beta - \alpha)P_1(x)P_2(x) & \text{if } x = y, \\ \alpha - (\alpha - \beta)P_2(y) & \text{if } x > y. \end{cases}$$

In order to determine whether the game has a saddle-point, it is necessary to evaluate $\max_x \min_y M(x, y)$ and $\min_y \max_x M(x, y)$. From the fact that $P_1(x)$ and $P_2(y)$ are monotonic increasing functions, it follows that

$$\max_x \min_y M(x, y) = \max_x \min [(\alpha - \beta)P_1(x) + \beta, \alpha P_1(x) + \beta P_2(x) + (\gamma - \beta - \alpha)P_1(x)P_2(x), a - (\alpha - \beta)P_2(x)],$$

$$\min_y \max_x M(x, y) = \min_y \max [(\alpha - \beta)P_1(y) + \beta, \alpha P_1(y) + \beta P_2(y) + (\gamma - \beta - \alpha)P_1(y)P_2(y), a - (\alpha - \beta)P_2(y)].$$

The functions $(\alpha - \beta)P_1(x) + \beta$ and $\alpha - (\alpha - \beta)P_2(x)$ are monotonic increasing and monotonic decreasing, respectively, having a common value

$$z_0 = (\alpha - \beta)P_1(x_0) + \beta = \alpha - (\alpha - \beta)P_2(x_0)$$

for some x_0 for which $P_1(x_0) + P_2(x_0) = 1$. Suppose that $(\gamma - \beta - \alpha) > 0$. Then

$$\alpha P_1(x_0) + \beta P_2(x_0) + (\gamma - \beta - \alpha)P_1(x_0)P_2(x_0) > z_0.$$

Therefore,

$$\max_x \min_y M(x, y) = z_0 \quad \text{and} \quad \min_y \max_x M(x, y) > z_0.$$

Thus, if $(\gamma - \beta - \alpha) > 0$, the game does not have a saddle-point, and therefore no pure strategy solution for both sides.

If $(\gamma - \beta - \alpha) < 0$, then it follows that

$$\min_y \max_x M(x, y) = z_0, \qquad \max_x \min_y M(x, y) < z_0,$$

and the game again does not have pure strategy solutions.

The graph in Fig. 17 illustrates that $M(x, y)$ does not have a saddle-point. We assume that $\gamma - \beta - \alpha > 0$. The point P is defined by

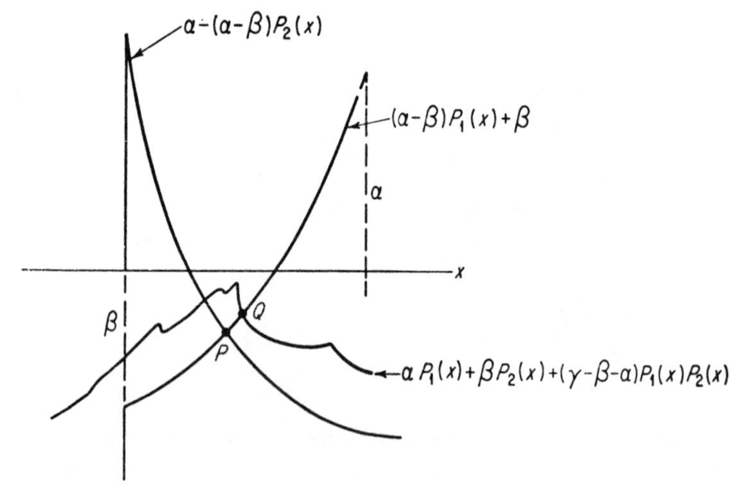

FIGURE 17

$P_1(x) + P_2(x) = 1$ and gives the $\max_x \min_y M(x, y)$. The point Q which cannot coincide with point P yields the $\min_y \max_x M(x, y)$.

In the special case where $\gamma - \beta - \alpha = 0$, the game does have a saddle-point at x_0 satisfying $P_1(x_0) + P_2(x_0) = 1$. This case was considered in the last section.

We summarize, without proof, the results of this type of duel. If the worths are such that $(\gamma - \beta - \alpha) > 0$, then the game has a value which is given by

$$v = (\alpha - \beta)P_1(x_0) + \beta,$$

where x_0 is such that

$$P_1(x_0) + P_2(x_0) = 1.$$

Blue has an optimal pure strategy

$$F^*(x) = I_{x_0}(x).$$

However, Red has no optimal strategy. He wishes to play a pure strategy as close as possible to x_0 but not equal to x_0.

If the worths are such that $(\gamma - \beta - \alpha) < 0$, then Red has an optimal pure strategy and Blue has no optimal strategy.

4. NOISY DUEL: MANY BULLETS, EQUAL ACCURACIES

If the duelists have more than one bullet each, and if we assume that the worths of the duelists are the same, namely, $+1$ to the surviving duelist and 0 otherwise, then it is relatively easy to compute the optimal strategy. Suppose that at time t, Blue and Red have $m(t)$ and $n(t)$ bullets, respectively, and equal accuracy functions, $p(t)$. An optimal strategy for either player is to fire one bullet whenever

$$p(t) = \frac{1}{m(t) + n(t)}.$$

However, the duelist with fewer bullets, at any such time, should not fire until

$$p(t) > \frac{1}{m(t) + n(t)}.$$

That is, he should hold his fire momentarily and then shoot only if his opponent did not shoot. The value of the game is

$$v = \frac{m(0) - n(0)}{m(0) + n(0)}.$$

For example, if Blue has two bullets and Red has three bullets and both players use their optimal strategies, then Red fires when the accuracy $p(t) = \frac{1}{5}$, both fire at $p(t) = \frac{1}{4}$, and both fire at $p(t) = \frac{1}{2}$, and $v = -\frac{1}{5}$ to Blue. In particular, if $p(t) = t$, then Red fires at $t = \frac{1}{5}$, both fire at $t = \frac{1}{4}$, and both fire at $t = \frac{1}{2}$.

5. NOISY DUEL: ONE BULLET, ARBITRARY ACCURACIES

Up to this point we have assumed that the accuracies of the duelists increase with time, and approach certainty. We now drop this restriction. Assume that the accuracies $p_1(t)$, $p_2(t)$ at time $t > 0$ are continuous functions, not necessarily monotonic, and that $p_1(t) = p_2(t) = C$ for all t larger than some t_0. We summarize without proof the optimal strategies of the duelists. If Blue fires at t_1 and misses, Red fires at t_2 for which $p_2(t)$ is a maximum in the interval t_1, t_0. For a pair of strategies (t_1, t_2), the payoff to Blue is

$$\begin{aligned} f(T) &= p_1(T) - [1 - p_1(T)]m_2(T) && \text{if } t_1 < t_2, \\ g(T) &= -p_2(T) + [1 - p_2(T)]m_1(T) && \text{if } t_2 < t_1, \\ h(T) &= p_1(T) - p_2(T) && \text{if } t_1 = t_2, \end{aligned}$$

where $T = \min(t_1, t_2)$, $m(T) = \max\limits_{t \geq T} p(t)$. Define

$$F(t) = \max_{\tau \leq t} f(\tau), \quad \text{and} \quad G(t) = \min_{\tau \leq t} g(\tau);$$

then

(i) If $F(t)$ and $G(t)$ intersect first at $T_0 > 0$, then $v = F(T_0) = G(T_0)$. There exist $t_1 \leq T_0$ and $t_2 \leq T_0$ such that $f(t_1) = g(t_2) = v$ and $\max(t_1, t_2) = T_0$. Approximately optimal strategies for Blue and Red are obtained by choosing random times near t_1 and t_2, respectively, thus eliminating any effect of the function $h(T)$.

(ii) If $F(0) \geq G(0)$, then $v = \text{med}\,\{f(0), g(0), h(0)\}$. If ϵ is a small positive number chosen at random, the solution (t_1, t_2) is $(0, \epsilon)$, $(\epsilon, 0)$, or $(0, 0)$, depending on whether the median is assumed at $f(0)$, $g(0)$, or $h(0)$, respectively.

Example. Suppose t_0 and the accuracies are

$$t_0 = \tfrac{2}{3}, \quad C = \tfrac{1}{3}, \quad p_1(t) = |\tfrac{1}{3} - t|, \; p_2(t) = \tfrac{1}{2} - |\tfrac{1}{2} - t|.$$

This is case 1, and $v = 0$, $t_1 = 0$, $t_2 = \frac{2}{5}$. If the time $t = \frac{2}{5}$ has arrived and Blue has erred by not firing, Red should re-evaluate on the basis of the new game beginning at $t = \frac{2}{5}$. This duel is Case 2 and its value is $-\frac{1}{6}$, $t_2 = \frac{1}{2}$, $t_1 = 0.54$.

6. SILENT DUEL: ONE BULLET EACH DUELIST, EQUAL ACCURACIES

In a silent duel each duelist knows how many bullets the duelists have at the start of the duel, but each duelist is ignorant of any firing by the other. Let us assume that each duelist has one bullet at the beginning of the duel. The duelists approach each other with no opportunity for retreat. Let us assume that the two duelists have the same accuracy functions; i.e., if the time of firing is the same, the duelists have the same probability of a kill.

Since the accuracies increase with time, each duelist wishes to postpone firing as long as possible. However, by postponing his firing he also increases the probability of being hit by his opponent. Further, the duel being silent, he does not know whether his opponent has fired. This conflict of interests may be resolved within the framework of game theory.

Let us assume that the worths of the duelists are the same, and assign to Blue the following values to the possible outcome:

$+1$, if Blue only survives,
-1, if Red only survives,
0, if both survive or neither survives.

Suppose that Blue fires when his accuracy is x and that Red fires when his accuracy is y, where $0 \leq x \leq 1$, $0 \leq y \leq 1$; then the payoff to Blue is

$$(9.1) \quad M(x, y) = \begin{cases} x + (1 - x)y(-1) = -y + (1 + y)x & \text{if } x < y, \\ x(1 - x) + x(1 - x)(-1) = 0 & \text{if } x = y, \\ y(-1) + (1 - y)x = -y + (1 - y)x & \text{if } y < x. \end{cases}$$

We find that

$$\max_x \min_y M(x, y) = 2\sqrt{2} - 3,$$

$$\min_y \max_x M(x, y) = 3 - 2\sqrt{2}.$$

Therefore the game does not have a saddle-point. Suppose that Blue uses a mixed strategy F and Red uses a pure strategy y; then Blue's expectation is

$$\begin{aligned} E(F, y) &= \int_0^1 M(x, y)\, dF(x) = \int_{x=0}^{x=y-0} M(x, y)\, dF(x) \\ &\qquad + \int_{y+0}^1 M(x, y)\, dF(x) \\ &= \int_0^{y-0} [-y + (1 + y)x]\, dF(x) \\ &\qquad + \int_{y+0}^1 [-y + (1 - y)x]\, dF(x) \\ &= -yF(y - 0) + (1 + y)\int_0^{y-0} x\, dF(x) - y + yF(y + 0) \\ &\qquad + (1 - y) \int_{y+0}^1 x\, dF(x) \qquad (9.2) \\ &= y[F(y) - F(y - 0)] - y + (1 + y) \int_0^{y-0} x\, dF(x) \\ &\qquad + (1 - y)\int_{y+0}^1 x\, dF(x). \end{aligned}$$

Suppose that the mixed strategy F is a density function over the interval $(\alpha, 1)$, or, i.e., $dF(x) = P(x)\, dx$ if $\alpha \leq x \leq 1$ and $dF(x) = 0$ if $x < \alpha$. Then

$$(9.3) \qquad E(F, y) = \begin{cases} -y + (1 + y) \int_\alpha^y xP(x)\, dx \\ \quad + (1 - y) \int_y^1 xP(x)\, dx & \text{if } y \geq \alpha \\ -y + (1 - y) \int_\alpha^1 xP(x)\, dx & \text{if } y \leq \alpha. \end{cases}$$

Since the game is symmetric, the value of the game is zero. It follows that, if the game has a density function P for a solution, then for all y for which $P(y) \neq 0$, $E(F, y) = v = 0$. Hence, if the game has a density function for a solution, we must have

$$-y + (1 + y) \int_\alpha^y xP(x)\,dx + (1 - y) \int_y^1 xP(x)\,dx = 0, \tag{9.4}$$

independent of y.

Differentiating the preceding expression twice with respect to y, we obtain

$$3P(y) + yP'(y) = 0.$$

Now, solving this differential equation, we get

$$P(y) = \frac{C}{y^3}. \tag{9.5}$$

By substituting (9.5) into (9.4), we have

$$-y + (1 + y)C \int_\alpha^y \frac{dx}{x^2} + (1 - y)C \int_y^1 \frac{dx}{x^2} = 0.$$

Hence

$$-y + C(1 + y)\left(\frac{1}{\alpha} - \frac{1}{y}\right) + C(1 - y)\left(\frac{1}{y} - 1\right) = 0,$$

independent of y for $\alpha \leq y \leq 1$. This yields

$$\alpha = \tfrac{1}{3}, \quad C = \tfrac{1}{4}.$$

Now we shall show that an optimal strategy for Blue and Red is to fire when the accuracy is x with the density function $P(x) = 1/4x^3$ where $\frac{1}{3} \leq x \leq 1$ and not to fire before $x = \frac{1}{3}$. Suppose Blue uses this strategy; then for $\frac{1}{3} \leq y \leq 1$, we have

$$E(F^*, y) = -y + \frac{(1 + y)}{4} \int_{1/3}^y \frac{dx}{x^2} + \frac{(1 - y)}{4} \int_y^1 \frac{dx}{x^2} = 0.$$

For $y \leq \frac{1}{3}$, we have

$$E(F^*, y) = -y + \frac{(1 - y)}{4} \int_{1/3}^1 \frac{dx}{x^2}$$

$$= -\frac{3}{2}y + \frac{1}{2} \geq 0.$$

Therefore the optimal strategy for Blue and Red is the following mixed strategy:

$$F^*(x) = 0 \qquad \text{if } 0 \leq x \leq \frac{1}{3},$$

$$= \frac{1}{8}\left(9 - \frac{1}{x^2}\right) \qquad \text{if } \frac{1}{3} \leq x \leq 1.$$

7. SILENT–NOISY DUEL: ONE BULLET EACH DUELIST

Mixing the duel, i.e., letting one duelist be informed of his opponent's firing, if any, yields optimal strategies which are different from the noisy or silent duels. Suppose that Blue is the silent duelist and that Red is the noisy duelist. Assume that the two duelists have the same accuracies. Then, if x and y are the accuracies at the time of firing for Blue and Red, respectively, the payoff to Blue is

$$M(x, y) = \begin{cases} x - y + xy & \text{for } x < y, \\ 1 - 2y & \text{for } x > y, \\ 0 & \text{for } x = y. \end{cases}$$

Let $a = \sqrt{6} - 2$; then it can be verified that the value of the game is

$$v = 1 - 2a = 0.101.$$

It can be verified that Blue, who has a silent bullet, has a unique optimal strategy, namely, the density function

$$f(x) = \begin{cases} 0 & \text{for } 0 \leq x < a, \\ \dfrac{\sqrt{2}\, a}{(x^2 + 2x - 1)^{3/2}} & \text{for } a \leq x \leq 1. \end{cases}$$

The optimal strategy for Red is also unique and is the same density $f(y)$ mixed with the pure strategy $y = 1$ in the ratio $2/a$. In terms of a cumulative distribution function, $G(y)$, the optimal strategy for Red is

$$G(y) = \frac{2}{2 + a} \int_0^y f(y)\, dy + \frac{a}{2 + a} I_1(y).$$

8. SILENT DUEL: ONE BULLET VERSUS TWO, EQUAL ACCURACIES

Giving one of the duelists an additional bullet considerably complicates the solution of the duel, even for the simple case where all accuracies are equal and monotonically increasing from zero to one. For safest results, the duelist with two bullets should shoot his two bullets in separate intervals but with a positive probability of saving the second bullet until the end. The duelist with one bullet defends himself against each of his opponent's bullets with two different density laws, spending slightly more than half of his fire-probability on the first.

Let the duelist with one bullet be Blue, who chooses a firing time x, with $0 \leq x \leq 1$. Red, who has two bullets, chooses two firing times, y and z, with $0 \leq y \leq z \leq 1$, where the time is identified with accuracy. The payoff to Blue is

$$M(x, y, z) = \begin{cases} x - (1 - x)y - (1 - x)(1 - y)z & \text{for } x < y \le z, \\ -y + (1 - y)x - (1 - y)(1 - x)z & \text{for } y < x \le z, \\ -y - (1 - y)z + (1 - y)(1 - z)x & \text{for } y \le z < x. \end{cases}$$

The solution of this duel is long and involved. We shall summarize, without proof, the optimal strategies in this duel. First, the game has a value. The duelist with one bullet, Blue, is at a disadvantage and the value of the duel to him is

$$v = \frac{2 - 3a}{2 + 3a}, \qquad \text{where } a = \sqrt{1 + \sqrt{\tfrac{1}{3}}} = 1.25593,$$
$$= -0.30650.$$

Blue also has a unique optimal mixed strategy, which is described by the density function

$$f(x) = \begin{cases} 0 & \text{if } 0 \le x \le \alpha, \\ \dfrac{k}{x^3} & \text{if } \alpha \le x \le b, \\ \dfrac{l}{x^3} & \text{if } b \le x \le 1, \end{cases}$$

where the constants are defined by

$$k = \frac{a(1 - b)}{1 + 2a - b} = 0.13805, \quad l = \frac{a}{1 + 2a - b} = 0.25760,$$

$$\alpha = \frac{1}{1 + 2a} = 0.28475, \qquad b = \frac{1}{1 + 2\sqrt{\tfrac{1}{3}}} = 0.46410.$$

If Blue uses this strategy, Red is forced to fire his first bullet between a and b, and his second bullet after b, if he is to ensure himself the value of the game. Red should fire his first bullet with probabilities described by the density function

$$g(y) = \begin{cases} 0 & \text{if } 0 \le y \le \alpha, \\ \dfrac{m}{y^3} & \text{if } \alpha \le y \le b, \\ 0 & \text{if } b \le y \le 1, \end{cases}$$

where $m = \dfrac{3}{4 + 6a} = 0.26006.$

The second bullet should be fired with a combination of a density function and a single step, as follows:

$$h(z) = \begin{cases} 0 & \text{if } 0 \leq z \leq b, \\ \dfrac{n}{z^3} & \quad b \leq z \leq 1, \\ \gamma & \quad z = 1, \end{cases}$$

where

$$\gamma = 2 - \sqrt{3} = 0.26795,$$

$$n = \frac{3\gamma}{2} = 0.40192.$$

The randomizations on y and z are carried out independently.

Figure 18 shows graphically the optimal strategies of the two duelists. The duelist with two bullets has a positive probability of saving his second bullet until the end. This is shown by the area of the wedge at $y = 1.0$.

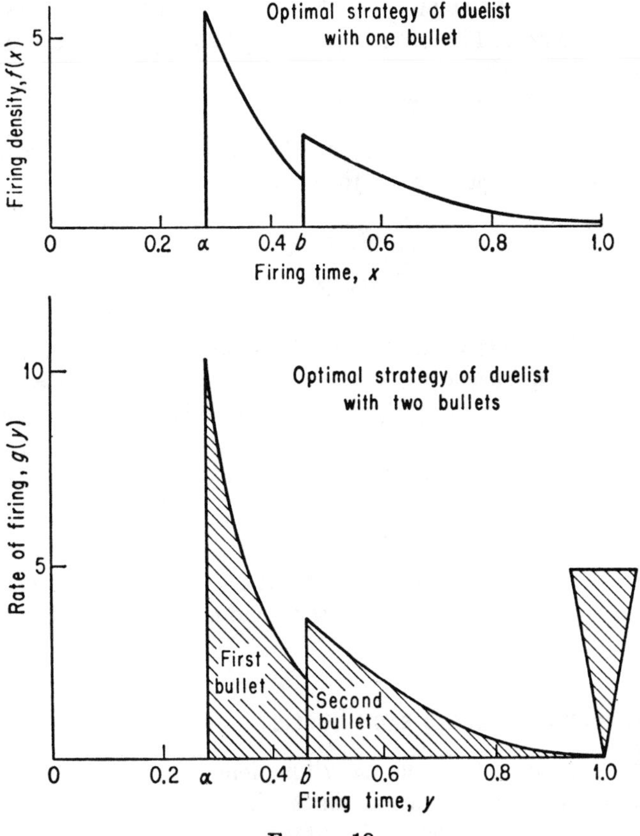

FIGURE 18

9. SILENT DUEL: POSITIVE INITIAL ACCURACY

Suppose that the two duelists have one silent bullet apiece and that their accuracies are equal and increase monotonically, but from an initial value $a \geq 0$ to 1. This situation may arise if neither duelist is permitted to fire before his accuracy reaches the value a. The introduction of this constraint modifies the solution of the silent duel. We describe, without proof, the optimal behavior of the duelists.

Let $f(x) = \frac{1}{4}x^3$ be the probability density function, which is also the solution of the silent duel without constraints. The optimal behavior of the duelists will depend on the size of a, as follows:

(i) If $0 \leq a \leq \frac{1}{3}$, then fire according to the probability density $f(x)$ in the interval $\frac{1}{3} \leq x \leq 1$ and never fire before $x = \frac{1}{3}$.
(ii) If $\frac{1}{3} \leq a \leq \frac{1}{2}$, let $(\frac{1}{3}, b)$ be the interval in which $f(x)$ has center of mass at a. Fire after b according to $f(x)$, and concentrate the remaining firing probability at a.
(iii) If $\frac{1}{2} \leq a \leq 1$, fire always at a.

To illustrate the solution, suppose that $a = \frac{4}{9}$, which falls in Case 2. To determine b, we need to solve the equation

$$\frac{\int_{1/3}^{b} xf(x)\,dx}{\int_{1/3}^{b} f(x)\,dx} = \frac{4}{9},$$

where $f(x) = \frac{1}{4}x^3$. This yields $b = \frac{2}{3}$. Therefore the optimal strategy for the duelists is to fire according to the probability density $f(x) = \frac{1}{4}x^3$ from $x = \frac{2}{3}$ to $x = 1$. This uses up $\frac{5}{32}$ of the firing probability. The remaining firing probability, $\frac{27}{32}$, is concentrated at $a = \frac{4}{9}$.

It will be observed that the average firing accuracy in each of the three cases above is

$$\max\left(\tfrac{1}{2}, a\right).$$

This is exactly the optimal firing time for the associated noisy duel, where the duelists have equal accuracy functions and may fire after the accuracy exceeds a.

10. SILENT DUEL: *m* BULLETS EACH DUELIST

Assume equal monotonic accuracy functions with each duelist having m bullets. Assume that the duel is silent. Then the optimal strategy is to fire the $(m + 1 - k)$th bullet in the interval

$$\frac{1}{2k+1} \leq x \leq \frac{1}{2k-1}$$

according to the inverse cubic law $\frac{1}{4}kx^3$. The value of the game is zero.

It is interesting to note that the solution of the noisy duel with many bullets requires the shooting of the $(m + k - 1)$th bullet at the harmonic mean, $\frac{1}{2}k$, of the same interval.

11. SILENT DUEL: STRICTLY MONOTONIC ACCURACIES

We now drop the restriction of equal monotonic accuracy functions. We consider the silent duel in which the accuracies are arbitrary but strictly monotonic. Each duelist has only one bullet.

Let $p_1(t)$ and $p_2(t)$ be the probabilities of a hit by Blue and Red, respectively. Let $p_1(0) = p_2(0) = 0$ and $p_1(1) = 1$. Define

$$f(t) = \frac{p_2'(t)}{p_1(t)p_2^2(t)}, \qquad g(t) = \frac{p_1'(t)}{p_2(t)p_1^2(t)}.$$

Let a_1 and a_2 be defined by the following equations:

$$\int_{a_1}^{1} \frac{1 - p_1(t)}{2} f(t)\, dt = 1,$$

$$\int_{a_2}^{1} \frac{1 - p_2(t)}{2} g(t)\, dt = 1,$$

and let

$$a = \max(a_1, a_2).$$

We now summarize the optimal strategy in terms of these defined quantities. The optimal strategy for Blue, whose accuracy function is $p_1(t)$, is to fire his bullet according to the following cumulative distribution function:

$$F(t) = \begin{cases} 0 & \text{for } 0 \le t \le a, \\ \dfrac{\int_a^t f(t)\, dt}{\int_{a_1}^1 f(t)\, dt} & \text{for } a \le t \le 1, \\ 1 & \text{for } t = 1. \end{cases}$$

The optimal strategy for Red, whose accuracy function is $p_2(t)$, is given by a similar cumulative distribution function:

$$G(t) = \begin{cases} 0 & \text{for } 0 \le t \le a, \\ \dfrac{\int_a^t g(t)\, dt}{\int_{a_2}^1 g(t)\, dt} & \text{for } a \le t \le 1, \\ 1 & \text{for } t = 1. \end{cases}$$

In other words, Blue has a jump at 1 if $a_1 < a$.

The value of the game is

$$v = \begin{cases} \dfrac{1 - 3p_2(t)}{1 + p_2(a)} & \text{if } a = a_1, \\[2ex] \dfrac{-1 + 3p_1(a)}{1 + p_1(a)} & \text{if } a = a_2. \end{cases}$$

Example. Let Blue have an accuracy function $p_1(x) = x$ and Red have an arbitrary accuracy function $p_2(x) = p(x)$. Let α, β be the probabilities that Blue and Red, respectively, fire at the time when their accuracies are 1. Table 3 summarizes the solution for various forms of $p(x)$.

Table 3. SILENT DUEL, MONOTONIC ACCURACIES

$p(x)$	a	t	α	β
x	0.333	0	0	0
$\dfrac{2x^2}{1+x^2}$	0.409	0.1021	0	0.0764
$\dfrac{x(3-x)}{2(2-x)}$	0.372	0.0838	0.0063	0
$\dfrac{x}{2-x}$	0.414	0.1720	0	0
x^2	0.481	0.2481	0	0.0729
$\dfrac{2x^3}{4x^2 - 3x + 1}$	0.415	0.0280	0	0.1741

12. SILENT DUEL: CONTINUOUS FIRE

If a duelist is permitted to vary his rate of fire along a continuous scale between zero and one, then the silent duel with continuous fire has the property that every mixed strategy is dominated by a pure strategy.

Let $R(t)$ be the rate of fire and $A(t)$ be the accuracy density function, i.e., the probability of a kill by a unit amount of fire in a unit time at time t. Then the probability of the opponent's survival over the interval $(0, t)$ is

$$P(R, t) = \exp\left\{-\int_0^t A(u)R(u)\,du\right\},$$

where

$$0 \le R(t) \le 1, \qquad \int_0^1 R(t)\,dt = \beta \le 1,$$

and β is the total amount of ammunition.

If S is the set of all strategies R, then a mixed strategy is a distribution function F where $\int_S dF(R) = 1$. If the duelist is not killed by time t, the probability of his opponent's survival is

$$\phi(F, t) = \int_S \exp\left\{-\int_0^t A(u)R(u)\,du\right\} dF(R).$$

Corresponding to any mixed strategy F, there is a pure strategy

$$R_F = \int_S R\, dF(R)$$

which is uniformly better than F, independent of the opponent's action.

13. TARGET PREDICTION

A classic military problem is how best to aim at a mobile target which is deliberately maneuvering so as to confound prediction of its position. The mobile target may be a ship, plane, or infantryman. Their attacker may be a bomber, an antiaircraft gun, or a sniper, respectively. In each case, there is a time lag between the detection of the target and the arrival of the projectile.

Suppose a battleship in midocean is aware of an enemy bomber's presence, but the plane is too high for the battleship to take any offensive measures against the plane. However, the battleship can maneuver in order to confound the prediction of its position. The ship is interested only in not being hit. The plane has one bomb and let us assume that the bomber's aim is excellent, but there is a time lag between the release of the bomb and its detonation. That is, the bomber must aim at an anticipated position of the ship.

In order to gain insight into this difficult problem it is necessary to simplify the problem further by assuming the ocean to be one-dimensional and discrete. Let us assume that the battleship is located on one of a long row of points and that at each unit of time it moves either one unit to the left or one unit to the right. Let us assume that the time lag is 2 units or, what is the same thing, 2 moves. The payoff to the bomber is 1 if he hits the battleship, and zero otherwise.

A strategy for the battleship will depend on prior moves. Since the battleship's course more than two moves ago is known to his opponent, it is reasonable to suppose that this dependence will not reach very far back. Let us suppose that the choice depends on the previous move only. Now suppose that the battleship moves in the opposite direction with probability x and moves in the same direction with probability $1 - x$. At the end of 2 moves the battleship will be located at one of three positions. The probabilities associated with these three positions are

$$M = \begin{cases} (1 - x)^2 \\ x \\ x(1 - x). \end{cases}$$

From the description of the game it follows that M represents the expected payoff.

If we restrict ourselves to these strategies, then an optimal strategy for the battleship is a choice x which makes the maximum of these three probabilities a minimum. This occurs at

$$x = (1 - x)^2,$$

or at

$$x^* = \frac{3 - \sqrt{5}}{2}.$$

It can be shown that the value of the game is given by $x^* = (3 - \sqrt{5})/2$ and that x^* is also the optimal strategy for the battleship. Further, this optimal strategy is unique. On the other hand, it can be shown that the bomber does not possess an optimal strategy. However, the bomber has an *ϵ-optimal strategy*. That is, for any $\epsilon > 0$, there is a mixed strategy for the bomber which assures him of a hit with probability $\geq x^* - \epsilon$, but no strategy guarantees x^*.

10 TACTICAL AIR-WAR GAME

1. INTRODUCTION

The problem of optimal employment of tactical air forces in the various theater air tasks, like many other military questions, can be analyzed as a multimove game between two sides. In the formulation of the game we view the tactical air war as consisting of a series of strikes, which are the moves of the game. Each move is an allocation of resources among various tasks. Among the usual tasks are the following:

Counter air. These operations are against the enemy's theater air base complex and organization in order to destroy his aircraft, personnel, facilities, etc.

Air defense. These represent air-defense operations against the enemy's counter-air operations.

Ground support. The targets for ground-support operations are concentrations of enemy troops or fortified positions, attacked in order to help the ground forces in the battle area. This is accomplished by aerial delivery of fire power against the enemy ground targets. We also include interdiction, reconnaissance, and airlift in this ground-support task.

2. FORMULATION OF TACTICAL GAME

We assume that the tactical air-war game consists of a series of strikes or moves. Each strike consists of simultaneous counter-air, air-defense, and ground-support allocation by each side. Let us assume that at the start of the game Blue has p planes and Red has q planes. Let us look at a strike in the game, say the initial strike. Suppose that on this strike Blue dispatches

x planes on counter-air operations and u planes on air-defense operations, and the remaining amount, $p - x - u$ planes, on ground-support operations. Similarly, suppose that for his first strike Red allocates y planes to counter air, w planes to air defense, and the remaining number, $q - y - w$ planes, to support his ground forces. For this initial strike and for any future strikes, the above decisions are made by each side in ignorance of the allocation of the opposing side. It is assumed, however, that each side knows the number p and q of planes that both have.

Since Red allocates w planes to air defense we can expect a reduction in the number of Blue's planes that get through to counter-air targets. The number of interceptions by Red will be proportional to w, say cw, unless Blue's attacking planes are saturated. This proportionality constant, or defense potential, depends on the planes' characteristics and flying altitudes, and on their weapons' characteristics. Hence the number of Blue attacking planes that penetrate Red's defenses is $x - cw$ as long as cw is not larger than x. If cw is larger than x, no Blue aircraft will penetrate. Therefore the number of Blue attacking planes that penetrate Red's defenses is the larger of the two numbers $x - cw$ and 0, or

$$\text{number Blue planes penetrating} = \max(0, x - cw).$$

The objective of Blue's counter-air operations is to reduce the enemy's air force by dropping bombs on certain targets, and the number of aircraft destroyed will vary with the number of attacking planes that penetrate Red's defenses. Increasing the number of Blue's penetrating planes will diminish the enemy's air force. If we assume that each of Blue's penetrating planes can destroy b planes of the enemy, then Blue could destroy $b \max(0, x - cw)$ Red planes, if Red had this number of aircraft at risk at the time of Blue's counter-air operations. The proportionality constant b depends on the target as well as the aircraft characteristics.

Suppose that during this initial strike Red's air force is increased by s replacements. Suppose also that of the q planes that Red uses on his operations, aq of them survive antiaircraft fire and accidents. Now let us assume that Red has $s + aq$ planes at risk at the time of Blue's counter-air operations. Then we have

$$\begin{aligned}\text{Red planes left} &= \max\{0, aq + s - \min[s + aq, b\max(0, x - cw)]\}\\ (10.1)\qquad &= \max\{0, \max[0, aq + s - b\max(0, x - cw)]\}\\ &= \max[0, aq + s - b\max(0, x - cw)].\end{aligned}$$

In exactly the same manner we can analyze the effect of the initial strike on Blue's inventory. At the end of the initial strike,

$$(10.2)\qquad \text{Blue planes left} = \max|0, dp + r - e\max(0, y - fu)|,$$

with the constants d, e, f defined similarly to a, b, c, respectively.

3. PAYOFF OF TACTICAL GAME

Let us look at Blue's employment of theater air forces during the campaign. We assume that his objective is to assist the ground forces in the battle area, and the outcome of this assistance will vary with the number of planes he allocates to ground-support operations. We assume that it is possible to construct for Blue a payoff function, giving the payoff for each strike of the campaign, in terms of the distance advanced by the ground forces which is a function of the number of planes allocated to ground support. This function depends heavily on the characteristics of the ground-support targets—i.e., on the degree of concentration of troops, vehicles, and materiel, and on the fortification of positions.

Now if Blue's ground forces must advance while being subjected to Red's ground-support sorties, Blue's yield in ground support is reduced in accordance with the number of planes allocated by Red to close-support missions. If $q - y - w$ is the function that measures the distance gained by Red's ground forces, then the net advance of Blue's ground forces, if he allocates $p - x - u$ planes to ground support and Red allocates $q - y - w$ planes to ground support, can be written as

$$p - x - u - (q - y - w).$$

The foregoing expression represents the payoff to Blue for this one period or one strike. The payoff for the entire campaign of N strikes is the sum of these net yields for each of the N strikes, or

$$M = \sum^{N} [p - x - u - (q - y - w)]. \tag{10.3}$$

The problem faced by each side is now apparent. For example, Blue would like to allocate a large number of planes to ground-support missions and thereby increase his payoff at a given move, yet he would like to destroy the Red air force by means of counter-air operations in order to ensure that q is small, or zero, for subsequent moves. Further, if he does not provide for air defense he may suffer severe losses to his own air force if Red elects to mount a large counter-air strike. Each player has to take into account the future and the possibilities open to his opponent.

4. TWO TASKS—COUNTER AIR AND GROUND SUPPORT

Let us first study the special tactical air war model with two tasks, counter air and ground support, and omit the air-defense task. This is equivalent to assuming that the air-defense potentials c and f are zero. Equations (10.1) and (10.2) now read

$$q_1 = \max(0, aq - bx + s), \tag{10.4}$$

$$p_1 = \max(0, dp - ey + r), \tag{10.5}$$

and the payoff (10.3) reduces to

$$M = \sum^{N} [(p - x) - (q - y)]. \tag{10.6}$$

It will be convenient to number the moves from the end of the game—i.e., the nth move means n moves to the end of the game. The tactical air war game may now be described as follows. Let p_n and q_n be the forces possessed by Blue and Red, respectively, and known by both Blue and Red at the start of a game of n moves. Suppose that at the initial move of this n-move game Blue allocates x_n planes to counter air where $x_n \leq p_n$, and simultaneously Red allocates y_n planes to counter air where $y_n \leq q_n$. Then the forces possessed by Blue and Red for the $(n - 1)$-move game are given by

$$p_{n-1} = \max(0, dp_n - ey_n + r_n), \tag{10.7}$$

$$q_{n-1} = \max(0, aq_n - bx_n + s_n). \tag{10.8}$$

Choices x_{n-1} and y_{n-1} are then made for this $(n - 1)$-move game. This process is continued for the duration of the game of N moves. The payoff (10.6) now becomes

$$M = \sum_{n=1}^{N} [(p_n - x_n) - (q_n - y_n)]. \tag{10.9}$$

The game as described has a value and optimal strategies. We may expect the value of the game to depend on the number of moves and the size of initial forces. Define $v_i(p_i, q_i)$ to be the value of the game of i moves in which Blue has p_i forces and Red has q_i forces at the start of this i-move game. In order to obtain the optimal initial move of an n-move game, i.e., the optimal values of x_n and y_n, it is sufficient to solve the game whose payoff is given by

$$M_n(x_n, y_n) = p_n - x_n - (q_n - y_n) + v_{n-1}(p_{n-1}, q_{n-1}), \tag{10.10}$$

where p_{n-1} and q_{n-1} are given by (10.7) and (10.8), respectively. Equation (10.10) defines the payoff in a game of where, on the initial move of the n-move game, Blue chooses x_n and Red chooses y_n, and then both players play optimally for the remaining $(n - 1)$ moves of the game.

5. OPTIMAL TACTICS FOR TWO TASKS

From (10.10) it is clear that in order to solve the n-move game, we first need to solve the $(n - 1)$-move game, which in turn requires a solution of the $(n - 2)$-move game, etc. Therefore we shall solve the n-move game by starting at the last move of the game and working backwards to the first move.

Let $v_0(p_0, q_0) = 0$. Then from (10.10) we have a one-move game or $n = 1$,

$$M_1(x_1, y_1) = p_1 - x_1 - q_1 + y_1. \tag{10.11}$$

Clearly, the optimal values of x_1 and y_1 are $x_1^* = 0$, $y_1^* = 0$. Blue and Red should allocate no resources to counter air. That is, the optimal tactics for Blue and Red on the last move are to allocate all their resources, p_1 and q_1, respectively, to ground support. The value of the one-move game is given by

$$v_1(p_1, q_1) = p_1 - q_1. \tag{10.12}$$

Having solved the one-move game, we now proceed to solve the two-move game. The payoff for $n = 2$ is given by

$$\begin{aligned} M_2(x_2, y_2) &= p_2 - x_2 - q_2 + y_2 + v_1(p_1, q_1) \\ &= p_2 - x_2 - q_2 + y_2 + p_1 - q_1, \end{aligned} \tag{10.13}$$

where p_1 and q_1 satisfy the following equations:

$$p_1 = \max(0, dp_2 - ey_2 + r_2), \tag{10.14}$$

$$q_1 = \max(0, aq_2 - bx_2 + s_2). \tag{10.15}$$

Substituting these values of p_1 and q_1 into (10.13), we obtain

(10.16)

$$M_2(x_2, y_2) = \begin{cases} p_2 - q_2 - x_2 + y_2 & \text{if } x_2 \geq \dfrac{aq_2 + s_2}{b},\ y_2 \geq \dfrac{dp_2 + r_2}{e} \\ (1 + d)p_2 - q_2 - x_2 \\ \quad + y_2(1 - e) + r_2 & \text{if } x_2 \geq \dfrac{aq_2 + s_2}{b},\ y_2 \leq \dfrac{dp_2 + r_2}{e} \\ (1 + d)p_2 - (1 + a)q_2 \\ \quad - (1 - b)x_2 + (1 - e)y_2 & \text{if } x_2 \leq \dfrac{aq_2 + s_2}{b},\ y_2 \leq \dfrac{dp_2 + r_2}{e} \\ \quad + r_2 - s_2 \\ p_2 - (1 + a)q_2 \\ \quad - (1 - b)x_2 + y_2 - s_2 & \text{if } x_2 \leq \dfrac{aq_2 + s_2}{b},\ y_2 \geq \dfrac{dp_2 + r_2}{e}. \end{cases}$$

The solution of the above game whose payoff $M_2(x_2, y_2)$ is defined in (10.16) will depend on the values of the parameters a, b, d, and e. First, suppose that $b > 1$ and $e > 1$. Then the optimal tactics are easily computed, and are given by

$$x_2^* = \frac{aq_2 + s_2}{b}, \qquad y_2^* = \frac{dp_2 + r_2}{e}.$$

The value of the game is

$$v_2 = \left(1 + \frac{d}{e}\right)p_2 - \left(1 + \frac{a}{b}\right)q_2 + \frac{r_2}{e} - \frac{s_2}{b}. \tag{10.17}$$

If $b < 1$ and $e < 1$ in (10.16), then the optimal tactics are

$$x_2^* = 0, \qquad y_2^* = 0.$$

In this case, the value of the game is

$$v_2 = (1 + d)p_2 - (1 + a)q_2 + r_2 - s_2. \tag{10.18}$$

The optimal tactics, x_2^* and y_2^*, tell Blue and Red how they should allocate their forces among the two tasks on the next to the last strike. Of course, on the last strike of the campaign we have already shown that Blue and Red should allocate all their forces to ground support.

Having computed the optimal tactics for $n = 1$ and $n = 2$, we next proceed to compute the optimal tactics for $n = 3$ (three moves to the end of the game). We set $n = 3$ in (10.9), which gives us the payoff function

$$M_3(x_3, y_3) = p_3 - x_3 - q_3 + y_3 + v_2(p_2, q_2), \tag{10.19}$$

where

$$p_2 = \max(0, dp_3 - ey_3 + r_3),$$

$$q_2 = \max(0, aq_3 - bx_3 + s_3),$$

and $v_2(p_2, q_2)$ is given by (10.17) or (10.18), depending upon the parameters b and e. Solving the game whose payoff function is defined by (10.19), we can obtain the optimal tactics for a three-move game and the game value $v_3(p_3, q_3)$.

Proceeding in the above manner, we build up the solution of the game for any number of moves. It turns out that, independent of the parameters, initial conditions, and size of forces at a given move, both sides have optimal pure strategies. Although every strike by a player is made simultaneously with his opponent, nevertheless, a player never needs to randomize. Thus an optimal strategy for a player can be specified by giving, for each strike of the campaign, either the number of planes he allocates to counter-air operations or the number of planes he allocates to ground support. These optimal allocations depend on the attrition parameters a, b, d, e, and on the number of strikes remaining in the campaign.

Before we describe the optimal strategies obtained by the above procedure, let us introduce some notation. From Equation (10.17) it is clear that if Red had enough forces he could annihilate Blue's forces by allocating

$$y_n = \frac{dp_n + r_n}{e}$$

to counter-air operations. Any allocation exceeding this is clearly wasteful. Thus it is reasonable to call an allocation

$$y_n = \min\left(q_n, \frac{dp_n + r_n}{e}\right)$$

a *concentration on counter air*. We shall denote such an allocation by A. Similarly the allocation by Blue of

$$x_n = \min\left(p_n, \frac{aq_n + s_n}{b}\right)$$

is a concentration on counter air by Blue and will also be denoted by A.

If Blue's allocation to counter air is $x_n = 0$, and therefore all forces p_n are allocated to ground support, we shall denote this tactic by G. Similarly, if $y_n = 0$, then Red concentrates on ground support, and this is denoted by G.

Finally, consider an allocation which is neither A nor G. We shall denote this by the symbol (A, G). In such a tactic a player splits his forces between counter air and ground support, and does not concentrate on either task.

The optimal strategies depend on the attrition parameters. Let us first suppose that

$$a + b \geq 1 \quad \text{and} \quad d + e \geq 1.$$

Then it turns out that the optimal strategy for each player requires him to begin the campaign with a series of allocations A, and to end with a series of allocations G. The points at which the players shift from A to G will, in general, be different for Red and Blue.

The precise points of shift depend on the magnitudes of the attrition parameters, in the following manner. Let f be the largest integer for which the inequality

$$\frac{1}{e} - \frac{1 - d^f}{1 - d} > 0$$

holds, and let g be the largest integer for which the inequality

$$\frac{1}{b} - \frac{1 - a^g}{1 - a} \geq 0$$

holds. These integers f and g determine the strike at which the shift is made.

The optimal strategies obtained by the method described above are summarized in Table 4. In this table, the integer t denotes another shift point which is applicable in cases 2 and 3.

6. OPTIMAL TACTICS FOR THREE TASKS

We now return to the more general model with all three tasks—counter air, air defense, and ground support—present. We shall see that increasing the number of air tasks to three leads to substantial changes in the character of the optimal tactics.

In order to simplify the analysis we assume that Blue and Red have the same air-defense potential: each plane allocated to defense can prevent one

Table 4. Optimal Allocation of Forces Between Two Tasks

Case No.	Range of parameters	Player	Optimal allocation at move number n (counting from end of campaign)			
			$1 \leq n \leq \min(f+1, g+1)$	$\min(f+2, g+2) \leq n \leq t$	$n = t+1$	$t+2 \leq n \leq N$
1	$a+b \geq 1, d+e > 1$	Blue	G	G	A	A
	$f < g$	Red	G	A	A	A
	$g < f$	Blue	G	A	A	A
		Red	G	G	A	A
	$g = g$	Blue	G	A	A	A
		Red	G	A	A	A
2	$a+b \leq 1, d+e \geq 1$	Blue	G	G	A	(A, G)
	$(g = \infty)$	Red	G	A	A	A
3	$a+b \geq 1, d+e \leq 1$	Blue	G	A	A	A
	$(f = \infty)$	Red	G	G	A	(A, G)
4	$a+b \leq 1, d+e \leq 1$	Blue	G	G	G	G
	$(f = g = \infty)$	Red	G	G	G	G

attacking plane from reaching the target—that is, we assume that $c = f = 1$. We also assume that each attacking plane that penetrates the defense can destroy one plane in an airfield strike, or $b = e = 1$, and that losses due to aborts, accidents, and antiaircraft fire are negligible, or $a = d = 1$. Finally, we assume that replacements are absent, i.e., $r = s = 0$. Then the inventory of planes at the end of a strike will be, for Blue and Red, respectively,

$$p_1 = \max [0, p - \max (0, y - u)],$$

$$q_1 = \max [0, q - \max (0, x - w)].$$

If we derive the optimal strategies in the three-task model we find that they are different from the two-task model in the following two important ways: first, the optimal tactics depend upon the relative strengths of the two sides; second, optimal play requires one player to use a mixed strategy.

The techniques used to derive the optimal tactics are similar to those used for the two-task model, but are more complicated. We shall omit the derivation; however, the reader can verify the given results. We shall give a complete description of the optimal employment of forces in terms of the number of strikes remaining and the relative strengths of the two sides. In describing the optimal allocations we shall always assume that at the move in question Blue is the stronger side, that is to say, $p \geq q$. This is merely a convention to facilitate the description of optimal tactics, and is not meant to imply that the side which is the stronger at a given stage of the game will always remain the stronger for all subsequent moves. Of course, if a player is initially stronger and plays optimally, then he will remain the stronger throughout the game.

First, we give a qualitative description of the optimal strategies:

1. *Campaign ends with ground support.* The campaign always ends with a series of strikes on ground support—i.e., during the closing period of the campaign both Red and Blue concentrate all their forces on ground-support missions. In this terminal period both sides have the same optimal tactics, regardless of their resources. If we assume that $c = f = b = e = 1$, then this terminal period consists of only the last two strikes of the campaign.

2. *Blue* (*stronger*) *splits his forces.* At all times other than the closing phase of the campaign, Red and Blue have very different optimal tactics. During any of these early strikes, the stronger side, say Blue, has a pure strategy, i.e., there exists a best allocation of Blue's air force among the three air tasks. In this connection, there is a critical value (about 2.7) of the ratio of the Blue force to the Red force which governs Blue's allocation during the early period in the following manner: If the force ratio is less than this critical value, then the optimal allocation in the early period consists of splitting the stronger force between two tasks, counter air and air defense, and neglecting the ground-support task. The size of the split de-

pends on the relative strengths of the two forces and the number of strikes left in the campaign. However, if Blue's strength relative to Red's is greater than the critical value, then regardless of his strength, Blue should divide his force in a fixed way among the three tasks, counter air, air defense, and ground support. The number of aircraft allocated to each mission, however, is still dependent on the number of strikes remaining.

3. *Red (weaker) mixes his tactics and concentrates his forces.* The weaker combatant cannot use a single strategy, but must bluff during all the strikes other than those of the terminal phase. Unlike his opponent, the weaker combatant does not have a single allocation that is best. He must use a mixed strategy and gamble for high payoffs. If he is not too weak—i.e., if the force ratio is less than the critical value—then he concentrates his entire force either on counter air or on air defense; but which of these tasks receives the full effort is decided by some chance device. However, if Red is very weak (force ratio larger than the critical value), then he allocates his entire air force to any one of the three air tasks with the particular task again chosen at random. In other words, if a player is very weak relative to the opponent, then he takes a chance on an early payoff. Of course, to be most effective, he must bluff correctly—i.e., the random device should select the tasks with the proper relative frequencies.

4. *Mix and split the same tasks.* It is of interest to note that on each strike Red, the weaker side, bluffs with the same tasks that Blue uses in his allocation. Thus if Red is very weak he bluffs with each of the three tasks. However, if Red is moderately weak, he bluffs with two tasks—counter air or air defense—and Blue splits his forces between the same two tasks.

5. *Blue's defense decreases during campaign.* As noted above, prior to the closing phase of the campaign Blue splits his forces among his air tasks. The actual split is a function of the force sizes of Blue and Red and the number of strikes left in the campaign. However, as the campaign proceeds, the fraction of Blue's force allocated to air defense will decrease. At the same time, the fraction allocated by Blue to counter air will increase. During this time, the chance that Red will attack Blue also decreases, but the chance that Red will defend himself increases.

6. *Blue's defense in a long campaign.* In the early stages of a relatively long campaign, the stronger side defends itself against a concentrated attack by the weaker side. During this period, Blue dispatches on air defense a force of planes approximately the size of Red's entire force. You will recall that a particular value for the air-defense effectiveness was assumed.

Table 5 summarizes the optimal tactics for campaigns consisting of at most eight strikes. The tabulation gives the optimal allocation for each strike (where the strike number is defined by the number of strikes remaining in the campaign) as a function of the relative sizes of the forces at the

Table 5. Optimal Allocation of Forces Among Three Tasks

(Strong side having p forces and weak side having q forces, $p > q$)

Duration of campaign (no. of strikes remaining in campaign)	*Relative initial strengths of opponents (ratio of strong side to weak side)* p/q	*Optimal initial allocation by strong side (force size allocated to)*			*Optimal initial allocation by weak side (probability of concentrating forces on)*			*Value of game*
		Counter Air	Air Defense	Ground Support	Counter Air	Air Defense	Ground Support	
1	1.00 to ∞	0	0	p	0	0	1	$p - q$
2	1.00 to ∞	0	0	p	0	0	1	$2(p - q)$
3	1.00 to 2.00	q	0	$p - q$	0.50	0.50	0	$3(p - q)$
	2.00 to ∞	$1.5q$	$0.5q$	$p - 2q$	0.50	0.50	0	$3(p - q)$
4	1.00 to 2.33	$0.5p + 0.5q$	$0.5p - 0.5q$	0	0.50	0.50	0	$4.5(p - q)$
	2.33 to ∞	$1.67q$	$0.67q$	$p - 2.33$	0.33	0.33	0.33	$4.00p - 3.33q$
5	1.00 to 1.1	$0.41p + 0.59q$	$0.59p - 0.59q$	0	0.53	0.47	0	$6.35p - 6.35q$
	1.70 to 2.45	$0.55p + 0.36q$	$0.45p - 0.36q$	0	0.45	0.55	0	$5.82p - 5.45q$
	2.45 to ∞	$1.70q$	$0.75q$	$p - 2.45$	0.25	0.30	0.45	$5.00p - 3.45q$
6	1.00 to 1.44	$0.32p + 0.68q$	$0.68p - 0.68q$	0	0.56	0.44	0	$8.39p - 8.39q$
	1.44 to 1.78	$0.40p + 0.56q$	$0.60p - 0.56q$	0	0.52	0.48	0	$8.00p - 7.83q$
	1.78 to 2.51	$0.59p + 0.22q$	$0.41p - 0.22q$	0	0.41	0.59	0	$7.04p - 6.12q$
	2.51 to ∞	$1.71q$	$0.80q$	$p - 2.51$	0.20	0.29	0.51	$6.00p - 3.51q$
7	1.00 to 1.29	$0.25p + 0.75q$	$0.75p - 0.75q$	0	0.58	0.42	0	$10.50p - 10.50q$
	1.29 to 1.53	$0.29p + 0.70q$	$0.71p - 0.70q$	0	0.57	0.43	0	$10.26p - 10.19q$
	1.53 to 1.84	$0.41p + 0.51q$	$0.59p - 0.51q$	0	0.50	0.50	0	$9.54p - 9.09q$
	1.84 to 2.55	$0.63p + 0.11q$	$0.37p - 0.11q$	0	0.37	0.63	0	$8.21p - 6.64q$
	2.55 to ∞	$1.72q$	$0.84q$	$p - 2.55$	0.17	0.28	0.55	$7.00p - 3.55q$
8	1.00 to 1.25	$0.20p + 0.80q$	$0.80p - 0.80q$	0	0.60	0.40	0	$12.60p - 12.60q$
	1.25 to 1.40	$0.22p + 0.78q$	$0.78p - 0.78q$	0	0.59	0.41	0	$12.48p - 12.45q$
	1.40 to 1.59	$0.28p + 0.69q$	$0.72p - 0.69q$	0	0.56	0.44	0	$12.06p - 11.86q$
	1.59 to 1.88	$0.42p + 0.46q$	$0.58p - 0.46q$	0	0.49	0.51	0	$11.03p - 10.22q$
	1.88 to 2.58	$0.66p + 0.01q$	$0.34p - 0.01q$	0	0.34	0.66	0	$9.36p - 7.07q$
	2.58 to ∞	$1.72q$	$0.86q$	$p - 2.58$	0.14	0.28	0.58	$8.00p - 3.58q$

time of that strike. However, the value of the game, which is given in the last column of the tabulation, is for the campaign of given duration. Using the definitions of optimal strategy, one can verify the values given in the table.

11 INFINITE GAMES WITH SEPARABLE PAYOFF FUNCTIONS

1. INTRODUCTION

In Chapter 3 we described a general method or procedure for solving finite games. Thus, given a game with a finite number of strategies, we can compute all the optimal strategies of Blue and Red. Since the method requires inverting a finite number of submatrices, it is a finite process.

No general method of solution exists for infinite games. That is, given an arbitrary game with an infinite number of strategies, we do not have a finite process for obtaining the optimal strategies of Blue and Red. However, there are some special methods applicable to special games classified according to the form of the payoff function. We have already described in Chapter 7 a method of solution for a large class of games, those games whose payoff is a convex function of Red's strategies. We shall now discuss another large class of games, those whose payoff function is a separable function of Blue's and Red's strategies.

2. DEFINITION

Suppose Blue chooses a strategy x from the interval $0 \leq x \leq 1$ and Red chooses a strategy y from the interval $0 \leq y \leq 1$. If the payoff function has the form

$$M(x, y) = \sum_{j=1}^{n} \sum_{i=1}^{m} a_{ij} r_i(x) s_j(y)$$

where the functions r_i and s_j are continuous, then the game is called a *separable* game.

Suppose that $n \leq m$. Then by rewriting the payoff function as

$$M(x, y) = \sum_{j=1}^{n} \left[\sum_{i=1}^{m} a_{ij} r_i(x) \right] s_j(y),$$

we can represent the payoff function as a single sum

$$M(x, y) = \sum_{j=1}^{n} r_j'(x) s_j(y),$$

where $r_j'(x) = \sum_{i=1}^{m} a_{ij} r_i(x)$. However, we shall use the double-sum form of of the payoff function since it provides greater mathematical flexibility.

3. MOMENTS

A mixed strategy for Blue may be represented by a cumulative probability distribution function $F(x)$. A mixed strategy for Red is a cumulative probability distribution function, $G(y)$. The expectation, or mixed strategy payoff, for a separable game is then given by

$$\phi(F, G) = \int_0^1 \int_0^1 M(x, y)\, dF(x)\, dG(y)$$

$$= \sum_{j=1}^{n} \sum_{i=1}^{m} \int_0^1 r_i(x)\, dF(x) \int_0^1 s_j(y)\, dG(y).$$

Let

$$r_i = \int_0^1 r_i(x)\, dF(x), \qquad i = 1, 2, \ldots, m,$$

and

$$s_j = \int_0^1 s_j(y)\, dG(y), \qquad j = 1, 2, \ldots, n.$$

Then to each distribution function $F(x)$ there corresponds a vector $r = (r_1, r_2, \ldots, r_m)$. To each distribution function $G(y)$ there corresponds a vector $s = (s_1, s_2, \ldots, s_n)$. We shall refer to the vector r as the m *moments* of $F(x)$. Similarly the vector s represents the n moments of $G(y)$.

In terms of the components of these vectors the mixed strategy payoff is the bilinear form

$$E(r, s) = \sum_{j=1}^{n} \sum_{i=1}^{m} a_{ij} r_i s_j.$$

4. EQUIVALENCE OF $F(x)$ AND POINTS OF CONVEX SET R

Letting $F(x)$ vary over all distribution functions, we obtain a set R of points $r = (r_1, r_2, \ldots, r_m)$. We shall now show that this set is the convex set D spanned by the curve C traced out in m dimensions by

$$r_i = r_i(x), \qquad i = 1, 2, \ldots, m,$$

as x varies between 0 and 1.

Consider the convex set D spanned by C. Suppose $r^0 = (r_1^0, \ldots, r_m^0)$ is a point of R not in D and let $F^0(x)$ be a distribution function which yields r^0, or

$$r_i^0 = \int_0^1 r_i(x)\, dF^0(x) \qquad i = 1, 2, \ldots, m.$$

Then there exists some hyperplane separating r^0 from D. That is, for some fixed $\delta > 0$,

$$\sum_{i=1}^m c_i r_i^0 - \sum_{i=1}^m c_i r_i(x) \geq \delta$$

for all x in $0 \leq x \leq 1$.

Integrating both sides with respect to $dF^0(x)$, we obtain

$$\sum_{i=1}^m c_i r_i^0 \int_0^1 dF^0(x) - \sum_{i=1}^m c_i \int_0^1 r_i(x)\, dF^0(x) \geq \delta \int_0^1 dF^0(x),$$

or

$$\sum_{i=1}^m c_i r_i^0 - \sum_{i=1}^m c_i r_i^0 \geq \delta,$$

giving the contradiction $0 > 0$. This proves that all points of R are in D.

Conversely, we shall show that every point in D is in R. Suppose r^0 is some point in D. That is, the components of r^0 may be represented as

$$r_i^0 = \sum_{k=1}^m \alpha_k r_i(x_k), \qquad i = 1, 2, \ldots, m,$$

where $\alpha_k \geq 0$ and $\sum_{k=1}^m \alpha_k = 1$. It is easily seen that the distribution function

$$F^0(x) = \sum_{k=1}^m \alpha_k I_{x_k}(x)$$

yields the point r^0. Hence every point of D is in R.

In a similar manner we can show that if we allow $G(y)$ to vary over all distribution functions we obtain a set S of points $s = (s_1, s_2, \ldots, s_n)$ which is identical with the convex set spanned by the curve C' traced out in n dimensions by

$$s_j = s_j(y), \qquad j = 1, 2, \ldots, n,$$

as y varies between 0 and 1.

5. BILINEAR GAME OVER A CONVEX SET

We have shown that if an infinite game has a separable payoff function every mixed strategy $F(x)$ corresponds to some point r in a convex set R. Conversely, every point in R corresponds to at least one mixed strategy $F(x)$. Therefore, the selection of a mixed strategy is equivalent to the selection of a point in a convex set R in m-dimensional space.

A separable game having a payoff function

$$M(x, y) = \sum_{j=1}^{n} \sum_{i=1}^{m} a_{ij} r_i(x) s_j(y)$$

is thus equivalent to the following game: Blue chooses a point $r = (r_1, \ldots, r_m)$ of a convex set R lying in Euclidean m-space. Red chooses a point $s = (s_1, \ldots, s_n)$ of a convex set S in Euclidean n-space. The payoff from Red to Blue is given by

$$E(r, s) = \sum_{j=1}^{n} \sum_{i=1}^{m} a_{ij} r_i s_j.$$

Further, the set R is the convex set spanned by the curve defined by

$$r_i = r_i(x), \qquad i = 1, 2, \ldots, m,$$

as x varies between 0 and 1. The set S is the convex set spanned by the curve defined by

$$s_j = s_j(y), \qquad j = 1, 2, \ldots, n,$$

as y varies between 0 and 1.

Optimal strategies of the two players are given by points r^0, s^0 such that

$$\min_{s \varepsilon S} E(r^0, s) = \max_{r \varepsilon R} E(r, s^0) = v,$$

where v is the *value* of the game. From the minimax theorem for bilinear forms over convex sets, it follows that there exist optimal strategies for the players. Of course, the existence of optimal strategies $F^*(x)$, $G^*(y)$ for the original game in terms of the separable payoff function was guaranteed by the fact that $r_i(x)$ and $s_j(y)$ are continuous.

6. DISTRIBUTION FUNCTION $F(x)$ AND POINTS OF CONVEX SET R

We have seen that to each distribution function $F(x)$ there corresponds some point r of R, and conversely. Every point on the curve C has for its coordinates

$$r_i = r_i(t), \qquad i = 1, 2, \ldots, m; \; 0 \leq t \leq 1.$$

Now every one-step distribution function $F(x) = I_t(x)$, where $0 \leq t \leq 1$ corresponds to a point r of R such that

$$r_i = \int_0^1 r_i(x) \, dI_t(x) = r_i(t) \qquad i = 1, 2, \ldots, m.$$

Therefore points on C correspond to one-step distribution functions.

Every point in R can be represented as a convex combination of at most m points of C. It follows from above that every point of R corresponds to a step-function having at most m steps.

7. NUMBER OF STEPS IN STEP-FUNCTION SOLUTION OF GAME

Every separable payoff function may be expressed as

$$M(x, y) = \sum_{j=1}^{n} \left(\sum_{i=1}^{m} a_{ij} r_i(x) \right) s_j(y)$$

$$= \sum_{j=1}^{n} r'_j(x) s_j(y)$$

where $n \leq m$. This is a bilinear game played over the convex sets R', S determined by the curves $r'_j = r'_j(x)$ and $s_j = s_j(y)$, respectively. Since the dimensions of both R and S are n, R' and S are determined by step-functions with at most n steps. Therefore, each player has an optimal mixed strategy with at most min (m, n) steps.

Suppose the payoff function is further generalized and is given by

$$M(x, y) = \sum_{j=1}^{\infty} \sum_{i=1}^{m} a_{ij} r_i(x) s_j(y),$$

and the convergence is uniform in y. Then the functions

$$s'_i(y) = \sum_{j=1}^{\infty} a_{ij} s_j(y), \qquad i = 1, 2, \ldots, m,$$

are continuous. The payoff may therefore be rewritten

$$M(x, y) = \sum_{i=1}^{m} r_i(x) s'_i(y)$$

From this it follows again that each player has an optimal mixed strategy with at most m steps.

8. SOLUTION OF SEPARABLE GAMES

A solution of a separable game, or a pair of optimal strategies of the two players, is a pair of points r^* of R and s^* of S such that

$$\min_{s \varepsilon S} \sum_{i,j=1}^{m,n} a_{ij} r_i^* s_j = \max_{r \varepsilon R} \sum_{i,j=1}^{m,n} a_{ij} r_i s_j^* = v,$$

where r_i and s_j are the components of r, s, respectively. We shall show how to find r^*, s^*.

The payoff function for a separable game may be written as follows:

$$E(r, s) = \sum_{j=1}^{n} \left(\sum_{i=1}^{m} a_{ij} r_i \right) s_j = \sum_{i=1}^{m} \left(\sum_{j=1}^{n} a_{ij} s_j \right) r_i.$$

If we let

$$f_j(r) = \sum_{i=1}^{m} a_{ij}r_i \qquad j = 1, 2, \ldots, n,$$

$$g_i(s) = \sum_{j=1}^{n} a_{ij}s_j \qquad i = 1, 2, \ldots, m,$$

we have

$$E(r, s) = \sum_{j=1}^{n} f_j(r)s_j = \sum_{i=1}^{m} g_i(s)r_i.$$

The planes $f_j(r) = 0$ can be considered to divide the space R into a finite number of portions $R_1, \ldots, R_i, \ldots, R_l$. For each point r^0 of R_i consider the set of points $S(r^0)$ of S where $\min_S \Sigma f_j(r^0)s_j$ is assumed. This is a convex set generally lying on the boundary of S. Similarly, the planes $g_i(s) = 0$ divide the space S into a finite number of portions $S_1, \ldots, S_j, \ldots, S_p$. Consider, for each s^0 of S, the set of points $R(s^0)$ where $\max_R \Sigma g_i(x)r_i$ is reached. If a point r^0 has the property that $S(r^0)$ contains a point s^0 for which $R(s^0)$ contains r^0, then r^0 is an optimal strategy for Blue. A similar analysis can be applied to Red.

By considering each region R_i of $R_1, R_2, \ldots, R_l$ and using the above mapping of R into the space S and then mapping the regions of S into R, we can obtain all solutions of both players as the fixed points in these mappings.

Example. Consider the following separable payoff function:

$$M(x, y) = y^2\left(\cos\frac{\pi}{2}x + \sin\frac{\pi}{2}x - 1\right) + \frac{4y}{3}\left(\cos\frac{\pi}{2}x - 3\sin\frac{\pi}{2}x\right) + \frac{1}{3}\left(5\sin\frac{\pi}{2}x - 3\cos\frac{\pi}{2}x\right).$$

This function does not have a saddle-point, hence the game has a mixed strategy solution. Now the choice of a mixed strategy $F(x)$ by Blue is equivalent to his choosing a point $r = (r_1, r_2)$ in a set R which is the convex set spanned by the curve

$$r_1 = \sin\frac{\pi}{2}x, \qquad 0 \le x \le 1,$$

$$r_2 = \cos\frac{\pi}{2}x, \qquad 0 \le x \le 1.$$

Similarly, Red chooses a point $s = (s_1, s_2)$ in a set S which is the convex set spanned by the curve

$$s_1 = y, \qquad 0 \le y \le 1,$$

$$s_2 = y^2, \qquad 0 \le y \le 1.$$

The bilinear payoff becomes

$$E(r, s) = s_2(r_1 + r_2 - 1) + \tfrac{4}{3}s_1(r_2 - 3r_1) + \tfrac{1}{3}(5r_1 - 3r_2)$$
$$= r_1(s_2 - 4s_1 + \tfrac{5}{3}) + r_2(s_2 + \tfrac{4}{3}s_1 - 1) - s_2.$$

In Fig. 19, the convex set R is the set of points bounded by the circular arc ABC and the line ADC. The line BD, or $l_1 = \frac{4}{3}r_2 - 4r_1 = 0$, divides the space R into three regions—BD where $l_1 = 0$, DAB where $l_1 > 0$, and BCD where $l_1 < 0$.

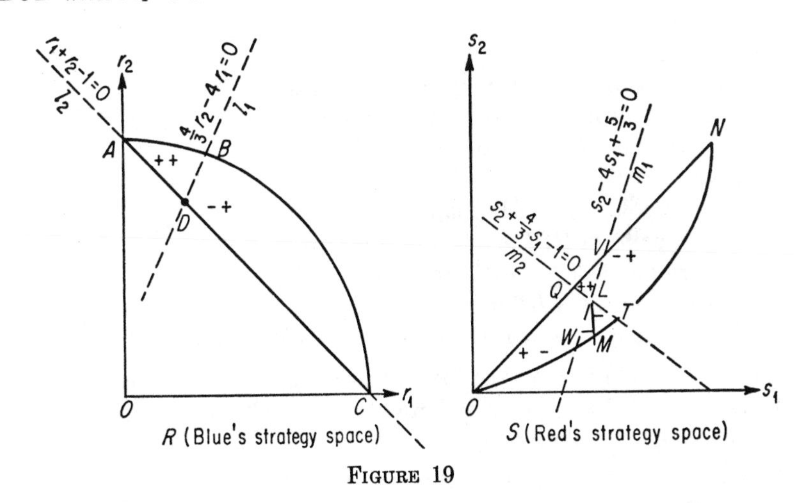

FIGURE 19

The convex set S is the set of points bounded by the parabolic arc OMN and the line OQN. The line VW, or $m_1 = s_2 - 4s_1 + \frac{5}{3} = 0$, the line QT or $m_2 = s_2 + \frac{4}{3}s_1 - 1 = 0$, and the line LM or $m_1 = m_2 \leq 0$, decompose S into five regions as follows:

$OQLW$, where	$m_1 \geq 0$,	$m_2 \leq 0$, excluding	$m_1 = m_2 = 0$;
QVL,	$m_1 \geq 0$,	$m_2 \geq 0$,	$m_1 = m_2 = 0$;
$LVNT$,	$m_1 \leq 0$,	$m_2 \geq 0$,	$m_1 = m_2 = 0$;
WLT,	$m_1 \leq 0$,	$m_2 \leq 0$,	$m_1 = m_2$;
LM,	$m_1 - m_2 \leq 0$.		

The set R is decomposed into three regions by means of $l_1 = 0$ and $l_2 = 0$, namely:

ABD, where	$l_1 \geq 0$,	$l_2 \geq 0$, excluding	$l_1 = l_2 = 0$;
BCD,	$l_1 \leq 0$,	$l_2 \leq 0$,	$l_1 = l_2 = 0$;
D,	$l_1 = l_2 = 0$.		

For every point r^0 in ABD, $E(r^0, s)$ assumes its minimum at $s_1 = s_2 = 0$.

However, for the point $s^0 = (0, 0) = 0$, $E(r, 0)$ assumes its maximum at $r = (1, 0)$ which is outside ABD. Therefore, no point in ABD can be an optimal strategy for Blue.

Now consider the region $LVNT$ of S. For each point s^0 of this region, $E(r, s^0)$ assumes its maximum at $r_2 = 1$, which is in region ABD. But region ABD, we have seen, maps into $s = 0$ which is in region $OQLW$, and not in $LVNT$. Thus no point of $LVNT$ can provide an optimal strategy.

We can summarize the complete mapping of the various regions of the two spaces as follows:

In Space R:

every point of ABD maps into O of space S;
each point of BCD maps into some point of $OWMN$ of S;
point D maps into every point of S.

In Space S:

every point of $OQLW$ maps into C of space R;
each point of QVL maps into some point of ABC or R;
every point of $LVNT$ maps into A of R;
each point of WLT maps into either A or C of R;
every point of LM maps into every point of ADC or R.

To find the solutions, or the fixed points, it is necessary to combine the two mappings. Letting $\longrightarrow$ stand for "map into," we can combine the above mappings as follows:

every point of $OQLW \longrightarrow C \longrightarrow N$;
each point of $QVL \longrightarrow$ some point of $ABC \longrightarrow O, OWMN$;
every point of $LVNT \longrightarrow A \longrightarrow O$;
each point of $WLT \longrightarrow A$ or $C \longrightarrow O$ or N;
every point of $LM \longrightarrow$ every point of $ADC = AD + DC + D \longrightarrow O, N, S$.

It is seen that only the last of the five regions yields fixed points. For, in this case we have, in particular,

$$LM \longrightarrow D \longrightarrow LM.$$

The solutions of the game are therefore:

for Blue: D: $r_1 = \frac{1}{4}$, $r_2 = \frac{3}{4}$;

for Red: LM: $s_1 = \frac{1}{2}$, $s_2 \leq \frac{1}{3}$.

The value of the game is $v = -\frac{1}{3}$.

We can express the solutions in terms of cumulative distribution functions as follows:

$$F^*(x) = \tfrac{3}{4} I_0(x) + \tfrac{1}{4} I_1(x),$$

$$G^*(y) = \left(1 - \frac{1}{2t}\right) I_0(y) + \frac{1}{2t} I_t(y), \qquad 0 < t \leq \tfrac{2}{3}.$$

Blue has a unique step-function solution, whereas Red has an infinite number of step-function solutions.

9. LOCAL DEFENSE OF TARGETS OF EQUAL VALUE

The solution to the problem of how to defend targets of equal value yields results which are not immediately evident. In defending such a system of targets, it is sometimes more profitable, depending on the payoff function, to leave some targets undefended and to defend others heavily. Although all the targets may have the same value, it is frequently a poor strategy to provide the same defense to each of them. In the model we shall study, it turns out that if the defender has a relatively small total force, then he never defends all targets but rather defends some heavily and leaves the remainder unprotected.

CASE 1. INDIVISIBLE DEFENSE.

Suppose the Defender has D units of defense to distribute among n targets of equal vlaue. Further suppose that the Defender may assign either 0, 1, or 2 units of defense to each target—i.e., the defense units are indivisible, and further that $D < 2n$. Assume that the Attacker has a total of A units, where $A < 2n$, with which to attack each target with either 0, 1, or 2 units.

We have an Attack-Defense game where both opponents know the parameters n, A, and D. A strategy for the Attacker is the choice of some number t,

$$0 \leq t \leq \frac{A}{2n}$$

representing the fraction of the targets subjected to attack by two units. Therefore, the fraction of targets attacked by single units is $A/n - 2t$. Of course, a strategy for the Attacker chooses only the number of targets to be attacked and leaves the selection of the specific targets to some random device. A strategy for the Defender is the choice of some number u, where

$$0 \leq u \leq \frac{D}{2n},$$

representing the fraction of targets to be defended by two units. Therefore $D/n - 2u$ is the fraction of targets having single defense units.

A possible payoff to the Attacker is the expected number of targets on which at least one bomb will be dropped. Now for each target, the expectation of its being destroyed will depend upon whether it is attacked by 1 or 2 units and defended by 0, 1, or 2 units. Thus for each target we have the payoffs shown in Table 6.

Table 6. Local Defense of Targets

(1) *Number attacking units*	(2) *Number defending units*	(3) *Probability of (1) and (2)*	(4) *Payoff if (1) and (2)*	(5) *Expected Payoff*
1	0	$\left(\frac{A}{n} - 2t\right)\left(1 - \frac{D}{n} + u\right)$	1	$\left(\frac{A}{n} - 2t\right)\left(1 - \frac{D}{n} + u\right)$
1	1	$\left(\frac{A}{n} - 2t\right)\left(\frac{D}{n} - 2u\right)$	0	0
1	2	$\left(\frac{A}{n} - 2t\right)u$	0	0
2	0	$t\left(1 - \frac{D}{n} + u\right)$	1	$t\left(1 - \frac{D}{n} + u\right)$
2	1	$t\left(\frac{D}{n} - 2u\right)$	$\frac{3}{4}$	$\frac{3}{4}t\left(\frac{D}{n} - 2u\right)$
2	2	tu	0	0

The payoff associated with n targets will be given by n times the sum of column (5) or

$$M(t, u) = n\left[\left(\frac{A}{n} - 2t\right)\left(1 - \frac{D}{n} + u\right) + t\left(1 - \frac{D}{n} + u\right) + \frac{3}{4}t\left(\frac{D}{n} - 2u\right)\right]$$

$$= -\frac{5}{2}ntu + \left(\frac{7}{4}D - n\right)t + Au + A - \frac{AD}{n}.$$

Since $M(t, u)$ is bilinear in t and u, it has a saddle-point, which is given by

$$t = \frac{2}{5}\frac{A}{n}, \quad u = \frac{2}{5}\left(\frac{7}{4}\frac{D}{n} - 1\right) \qquad \text{if } \frac{D}{n} > \frac{4}{7},$$

$$t = 0, \qquad u = 0, \qquad \text{if } \frac{D}{n} \leq \frac{4}{7}.$$

The value of the game is

$$v = \frac{3}{10}\frac{A}{n}(2n - D) \qquad \text{if } \frac{D}{n} > \frac{4}{7},$$

$$= \frac{A}{n}(n - D) \qquad \text{if } \frac{D}{n} \leq \frac{4}{7}.$$

For example, if the Defender has 100 defense units with which to defend 100 targets of equal value against an Attacker having a total of 100 units, an optimal strategy for the Defender is to place two defense units at each of 30 targets, one unit of defense at each of 40 targets, and leave 30 targets undefended. An optimal strategy for the Attacker is to use two attack units on each of 40 targets, one attack unit on each of 20 targets and no-attack on 40 targets. Of course, the specific targets to be defended or attacked are picked at random. The expected number of targets destroyed is 30.

If the Defender had placed one unit at each of the 100 targets, then the Attacker could destroy as many as 37.5 targets by attacking each of 50 targets with two units.

CASE 2. ARBITRARY DEFENSE.

If we assume that the Defender is able to distribute his forces continuously—i.e., he can place any part of a defense unit at a target—then the local defense problem becomes an infinite game with a separable payoff function. Again, the best defense strategy may require a heavy defense for some targets and no defense for others. As before, the Defender has D units of defense with which to defend n targets of equal value, where $D \leq 2n$. He is able to distribute continuously these D units among the n targets. The Attacker has A units of attack, where $A \leq 2n$, with which to attack some or all n targets. However, the Attacker may engage any given target with zero, one, or two units only. The payoff to the Attacker is measured by the expected number of targets destroyed.

A strategy for the Defender is a distribution function $G(y)$ representing the fraction of his targets defended by no more than y units of defense at each target. Since no target will be attacked by more than two units, it would be wasteful to have y larger than 2.

A strategy for the Attacker is some number t where $u = t \leq A/2n$, representing the fraction of targets to be attacked by two units. The selection of particular targets to be attacked and particular targets to be defended is made at random subject to the given fractions.

The payoff to the Attacker is the expected number of targets destroyed. The Attacker uses $2tn$ units of attack for double attacks on targets and the remainder, $A - 2tn$ units, for single attacks. Now consider a target which is defended by y units of defense and is under double attack. The probability that the target is destroyed is the probability that the two attacking units are not destroyed, or

$$1 - \frac{y}{2} \cdot \frac{y}{2} = 1 - \frac{y^2}{4}.$$

Now, the probability that a given target has no more than y units of defense is $G(y)$; it follows that the Attacker's *a priori* expectation of destroying any given target is

$$E_1 = \int_0^2 \left(1 - \frac{y^2}{4}\right) dG(y).$$

For the $2tn$ targets, the Attacker's expectation for his double attacks will be $2tnE_1$. Similarly, the Attacker's expectation for singly attacked targets is

$$(A - 2nt) \int_0^1 (1 - y)\, dG(y)$$

with the upper limit of integration extending only to 1, since no payoff is expected for targets defended by more than one unit. Therefore, the total payoff to the Attacker corresponding to each t and G will be

$$(11.1) \quad M(t, G) = 2nt \int_0^2 \left(1 - \frac{y^2}{4}\right) dG(y) + (A - 2nt) \int_0^1 (1 - y)\, dG(y).$$

The Defender must distribute all D units of defense among his n targets; hence he is restricted to pick only those G's for which

$$(11.2) \qquad n \int_0^2 y dG(y) = D.$$

We thus have a game in which the maximizing player picks some number t in the range $0 \leq t \leq A/2n$, and the minimizing player picks a distribution function satisfying (11.2), where the payoff, $M(t, G)$, to the Attacker is given by (11.1). There exist optimal strategies for the two players which will necessarily depend on the parameters $S = A/n$ and $T = D/n$, average attack per target and average defense per target. The solutions are obtained by analyzing the convex set spanned by the curve (μ_1, μ_2) defined by

$$\mu_1 = \int_0^2 \left(1 - \frac{y^2}{4}\right) dG(y) - 2 \int_0^1 (1 - y)\, dG(y),$$

$$\mu_2 = \int_0^2 y\, dG(y),$$

for all distribution functions $G(y)$ over the interval $(0, 2)$. In terms of this new set the game has a saddle-point which is readily obtained by the mapping method.

Table 7 presents a complete summary of the solutions to the game and their values for all possible values of average defense per target and average attack per target. The solutions are independent of the number of targets, n, but depend on the parameters S and T. One can easily verify the solutions shown in the table.

In region I defined by $0 \leq S \leq 1$ and $0 \leq T \leq 4 - 2\sqrt{3}$, the Defender may distribute his D units equally among the n targets, giving each target D/n units. In region II he may leave some targets undefended and defend the remaining targets equally. In region III the Defender must leave some targets undefended and place either one or two units of defense at the remaining number of targets. In region IV, the Defender places two units at each of the $D/2$ targets and the rest necessarily are undefended. In region V he places one unit at each of D targets leaving $n - D$ undefended. The Attacker has a pure strategy.

Table 7. Local Defense of Targets—Arbitrary Defense

Region No.	Range of		Optimal strategies		Value of game v
	$s = \frac{A}{n}$	$T = \frac{D}{n}$	Attack t^*	Defense G^*	
I	0 to 1	0 to $4 - 2\sqrt{3}$	0	Any G^* such that $\int_0^1 dG^*(y) = 1$	$(1 - T)S$
II	0 to 1	$4 - 2\sqrt{3}$ to $\frac{4}{7}$	0	Any G^* such that $\int_0^1 dG^*(y) = 1$, $\int y^2 \, dG^*(y) = 8T - 4$	$(1 - T)S$
III	0 to $\frac{5}{3}$	$\frac{4}{7}$ to 2	$\frac{2}{5}S$	$\frac{3(2 - T)}{10} I_0 + \frac{4 - 2T}{5} I_1 + \frac{7T - 4}{10} I_2$	$\frac{3}{10}(2 - t)S$
IV	$\frac{5}{3}$ to 2	0 to 2	$S - 1$	$\left(1 - \frac{T}{2}\right) I_0 + \frac{T}{2} I_2$	$1 - \frac{T}{2}$
V	1 to $\frac{5}{3}$	0 to $\frac{4}{7}$	$S - 1$	$(1 - T)I_0 + tI_1$	$\frac{4 - 7T + 3ST}{4}$

For example, if the Defender has 100 targets to defend, then, depending on the numbers of defense and attack units, the optimal strategies are given by Table 8.

10. SOLUTION OF POLYNOMIAL GAMES

If the payoff is a polynomial function of the two strategy variables x and y, we call this separable game a *polynomial game*. By introducing moments we transform the polynomial game to a bilinear game over a convex set, usually referred to as the *moment space*. The transformed game can be solved by mapping and determining the fixed points.

We shall describe the mapping method of solving polynomial games by illustrating it for a particular game. Suppose the payoff to Blue is given by

$$M(x, y) = 21x + 18x^2 - 24x^3 - 16y - 36xy - 9x^2y + 18x^3y + 60y^2 - 36y^3,$$

where $0 \leq x \leq 1$ and $0 \leq y \leq 1$ are pure strategies of Blue and Red, respectively. Let $F(x)$ and $G(y)$ be mixed strategies of Blue and Red. Define the following moments:

$$f_i = \int_0^1 x^i \, dF(x), \qquad i = 1, 2, 3,$$

$$g_j = \int_0^1 y^j \, dG(y), \qquad j = 1, 2, 3.$$

The choice of a mixed strategy $F(x)$ by Blue is equivalent to his choosing a point $f = (f_1, f_2, f_3)$ in a convex set R which is the convex hull of the curve C defined by

$$r_i = x^i, \qquad i = 1, 2, 3,$$

as x ranges from 0 to 1. Similarly, each mixed strategy $G(y)$ is associated with some point $g = (g_1, g_2, g_3)$ in a convex set S which is the convex hull of the curve D defined by

$$s_j = y^j \qquad j = 1, 2, 3,$$

as y ranges from 0 to 1.

In terms of points of convex sets R and S, the payoff becomes the bilinear form

$$M(f, g) = 21f_1 + 18f_2 - 24f_3 - 16g_1 - 36f_1g_1 - 9f_2g_1 + 18f_3g_1 + 60g_2 - 36g_3.$$

The convex set R is a three-dimensional volume whose boundary consists of the curve C, lines connecting the point (0, 0, 0) with each point of C, and lines connecting the point (1, 1, 1) with C. There are no planes which form the boundary of R. Of course, the boundary of S is the same as the boundary of R.

We now have the bilinear game in which Blue picks some point (f_1, f_2, f_3)

Table 8. LOCAL DEFENSE OF TARGETS—SOLUTION FOR 100 TARGETS

Number attack units A	*Number defense units D*	*Optimal strategies: $n = 100$*											*Value of game*
		No. Targets Attacked by			Target Defense								
		2 Units	1 Unit	0 Units	No. Targ.	No. per Targ.	No. Targ.	No. per Targ.	No. Targ.	No. per Targ.	No. Targ.	No. per Targ.	*Number targets destroyed*
100	50	0	50	50	100	.50	0	0	0	0	0	0	50
100	55	0	55	45	76	.73	24	0	0	1	0	2	45
100	100	40	20	40	0	..	30	0	40	1	30	2	30
175	100	75	25	0	0	..	50	0	0	1	50	2	50
150	50	50	50	0	0	..	50	0	50	1	0	2	31

in R and Red picks some point (g_1, g_2, g_3) in S. Let us rewrite the payoff as follows:

$$M(f, g)$$
$$= -16g_1 + 60g_2 - 36g_3 + 3f_1(7 - 12g_1) + 9f_2(2 - g_1) + 6f_3(-4 + 3g_1).$$

The maximum value of a linear form over a convex set is assumed on the boundary of the convex set. Therefore, for any point (g_1, g_2, g_3) chosen by Red, Blue, who wishes to maximize $M(f, g)$, will pick some point on C or some line joining (0, 0, 0) to C, or some line joining (1, 1, 1) to C. We shall first determine when Blue selects each of these three types of boundaries.

In order for Blue to select some line joining (0, 0, 0) to a point on C, it is necessary that

$$M(0, g) = M(f^0, g),$$

where f^0 is on C or $f_1^0 = x$, $f_2^0 = x^2$, $f_3^0 = x^3$. It is also necessary that

$$\frac{dM(f^0, g)}{dx} = 0$$

and

$$\frac{d^2M(f^0, g)}{dx^2} \leq 0.$$

The first two conditions yield the equations

$$3x(7 - 12g_1) + 9x^2(2 - g_1) + 6x^3(-4 + 3g_1) = 0,$$
$$3(7 - 12g_1) + 18x(2 - g_1) + 18x^2(-4 + 3g_1) = 0.$$

Solving these equations, we obtain

$$x = \tfrac{1}{2}, \qquad g_1 = \tfrac{2}{3}.$$

These values satisfy the necessary conditions for a maximum. Therefore, if Red chooses a mixed strategy for which $g_1 = \frac{2}{3}$, Blue can maximize $M(f, g)$ by choosing any point in R along the line joining (0, 0, 0) to $(\frac{1}{2}, \frac{1}{4}, \frac{1}{8})$. From this it follows that if Red chooses $g_1 > \frac{2}{3}$, then Blue will choose the point (0, 0, 0) in R, and if Red chooses $g_1 < \frac{2}{3}$, Blue will choose a point (x, x^2, x^3) in R satisfying the equation

$$\frac{dM(f, g)}{dx} = 3(7 - 12g_1) + 18x(2 - g_1) + 18x^2(-4 + 3g_1) = 0$$

or

$$x = \frac{-3(2 - g_1) - \sqrt{3(75g^2 - 150g_1 + 68)}}{6(3g_1 - 4)}.$$

We have partitioned the space S. We now need to partition the space R. Again, we make use of the fact that the minimum of a linear form over a convex set is assumed on the boundary. Therefore, for any point (f_1, f_2, f_3)

chosen by Blue, the minimum value of $M(f, g)$ will be assumed either on D, or on some line joining $(0, 0, 0)$ in S to D, or on some line joining $(1, 1, 1)$ in S to D.

Let us rewrite the payoff function as follows:

$$M(f, g) = 21f_1 + 18f_2 - 24f_3 + g_1(-16 - 36f_1 - 9f_2 + 18f_3) + 60g_2 - 36g_3.$$

We can show that the coefficient of g_1 satisfies the inequality

$$-43 \leq -16 - 36f_1 - 9f_2 + 18f_3 \leq -16.$$

The payoff function $M(f, g)$ will assume its minimum value along a line joining the point $(1, 1, 1)$ in S to D for all points in R satisfying the three conditions:

$$\text{(i)} \qquad M(f, 1) = M(f, g^0),$$

where g^0 is on D or $g_1^0 = y$, $g_2^0 = y^2$, $g_3^0 = y^3$,

$$\text{(ii)} \qquad \frac{dM(f, g^0)}{dy} = 0,$$

and

$$\text{(iii)} \qquad \frac{d^2M(f, g^0)}{dy^2} \geq 0.$$

The first two conditions yield the equations

$$(-16 - 36f_1 - 9f_2 + 18f_3) + 60 - 36 = y(-16 - 36f_1 - 9f_2 + 18f_3) + 60y^2 - 36y^3,$$

$$(-16 - 36f_1 - 9f_2 + 18f_3) + 120y - 108y^2 = 0.$$

Solving, we obtain

$$-16 - 36f_1 - 9f_2 + 18f_3 = -28 \quad \text{and} \quad y = \tfrac{1}{3}.$$

These values satisfy the conditions for a minimum. Therefore, for all points in R for which

$$-16 - 36f_1 - 9f_2 + 18f_3 = -28,$$

the minimum value of $M(f, g)$ is assumed at each point of the line joining $(1, 1, 1)$ to $(\frac{1}{3}, \frac{1}{9}, \frac{1}{27})$. It follows that for all points in R for which

$$-16 - 36f_1 - 9f_2 + 18f_3 < -28,$$

the minimum of $M(f, g)$ is assumed at $(1, 1, 1)$ in S, and for all points in R for which

$$-16 - 36f_1 - 9f_2 + 18f_3 > -28,$$

the minimum of $M(f, g)$ is assumed at some point (y, y^2, y^3) in S satisfying the equation

$$\frac{dM(f, g)}{dy} = -16 - 36f_1 - 9f_2 + 18f_3 + 120y - 108y^2 = 0,$$

or

$$y = \frac{10 - \sqrt{100 + 3(-16 - 36f_1 - 9f_2 + 18f_3)}}{18}.$$

Table 9 presents a summary of the partitioning and mappings of spaces

Table 9

$$M(x, y) = 21x + 18x^2 - 24x^3 - 16y - 36xy - 9x^2y + 18x^3y + 60y^2 - 36y^3$$

R-SPACE	
Points in R for which	*Map into points in S* (*location in S of* min (f, g))
$l_1 = -16 - 36f_1 - 9f_2 + 18f_3 < -28$	$(1, 1, 1)$
$l_1 = -17 - 36f_1 - 9f_2 + 18f_3 > -28$	$y = \frac{10 - \sqrt{100 + 3l_1}}{18}$
$l_1 = -16 - 36f_1 - 9f_2 + 18f_3 = -28$	$\beta(\frac{1}{3}, \frac{1}{9}, \frac{1}{27}) + (1 - \beta)(1, 1, 1)$ where $0 \leq \beta \leq 1$
S-SPACE	
Points in S for which	*Map into points in R* (*location in R of* max $M(f, g)$)
$g_1 > \frac{2}{3}$	$(0, 0, 0)$
$g_1 < \frac{2}{3}$	$x = \frac{-3(2 - g_1) - \sqrt{3(75g_1^2 - 150g_1 + 68)}}{6(3g_1 - 4)}$
$g_1 = \frac{2}{3}$	$\alpha(\frac{1}{2}, \frac{1}{4}, \frac{1}{8}) + (1 - \alpha)(0, 0, 0)$

R and *S*. The solution of the game, the optimal strategies, are the fixed points in the mappings listed in the table. We test each of the three regions in each space for the existence of fixed points. From the table we note that if $l_1 < -28$ then $M(f, g)$ assumes its minimum at $(1, 1, 1)$ in *S*, or $g_1 = 1$. Again from the table we note that if $g_1 = 1 > \frac{2}{3}$, then $M(f, g)$ assumes its maximum at $(0, 0, 0)$ in *S* or $l_1 = 16$; hence $l_1 > -28$. Therefore no point in *R* for which $l_1 < -28$ can provide a solution to the game. In a similar manner, we can show that $l_1 > -28$ and $g_1 < \frac{2}{3}$ contain no fixed points, and therefore no solutions of the game.

Since we have eliminated two of the three regions, it follows that $l_1 = -28$ and $g_1 = \frac{2}{3}$ must contain the fixed points. It remains to solve for the values of α and β. For the third region to contain a fixed point, we must have

$$\frac{\beta}{3} + 1 - \beta = \frac{2}{3}$$

$$-16 - 36\left(\frac{\alpha}{2}\right) - 9\left(\frac{\alpha}{4}\right) + 18\left(\frac{\alpha}{8}\right) = -28,$$

or

$$\alpha = \frac{2}{3}, \qquad \beta = \frac{1}{2}.$$

The fixed points are therefore $(\frac{1}{3}, \frac{1}{6}, \frac{1}{12})$ in R and $(\frac{2}{3}, \frac{5}{9}, \frac{14}{27})$ in S. If Blue chooses a strategy such that $f_1 = \frac{1}{3}$, $f_2 = \frac{1}{6}$, $f_3 = \frac{1}{12}$, then Red can minimize $M(f, g)$ by choosing $g_1 = \frac{2}{3}$, $g_2 = \frac{5}{9}$, $g_3 = \frac{14}{27}$. If Red chooses the preceding set of g's, then Blue can maximize $M(f, g)$ by choosing $f_1 = \frac{1}{3}$, $f_2 = \frac{1}{6}$, $f_3 = \frac{1}{12}$. This verifies that these points are fixed points and therefore a solution of the game. The solution can be expressed by

$$F^*(x) = \tfrac{1}{3}I_0(x) + \tfrac{2}{3}I_{1/2}(x),$$

$$G^*(y) = \tfrac{1}{2}I_{1/3}(y) + \tfrac{1}{2}I_1(y).$$

The value of the game is given by

$$v = \int_0^1 \int_0^1 M(x, y)\, dF^*(x)\, dG^*(y) = 6.$$

11. TACTICAL RECONNAISSANCE—SINGLE MISSION

In planning an attack on a target whose military worth is not known exactly, it may be advisable to invest in reconnaissance before the mission to determine the exact worth of the target. The uncertainty regarding the worth of the target may arise from unknown or partially known results of earlier strikes on the same target. If the exact worth of the target is discovered through reconnaissance, then it is possible to dispatch the most efficient size attacking force against it—i.e., a large force will not be committed against a worthless target and a small force will not be sent to a valuable target. Since reconnaissance is costly, it is necessary to evaluate its desirability against the potential outcomes of the attack.

A successful reconnaissance is assumed to reveal the exact value of the target. In order to be successful, at least one reconnaissance aircraft must fly to the target and return.

Let us introduce the following notation:

B = military worth of one bomber.
R = military worth of one reconnaissance aircraft.
T = military worth of the target.
$\phi(t)$ = probability that the value of the target does not exceed t. This probability distribution is known prior to the reconnaissance.

r = number of reconnaissance aircraft sent out prior to the mission.
b = number of bombers dispatched to the target during mission.
p = one-way survival probability of bomber and reconnaissance aircraft between base and target.
αT = probable worth of the target after being hit by one bomber.
$\alpha^2 T$ = probable worth of the target after being hit by two bombers.

The object of the attacker is to maximize the net outcome of the mission, namely the difference between the target damage and the aircraft losses. Thus the payoff will depend on r and b and is given by

$$M(r, b) = \int [t(1 - \alpha^{pb}) - (1 - p^2)Bb - (1 - p^2)Rr] \, d\phi(t),$$

where the first term under the integral represents the target damage and the remaining two terms are the worths of the bomber and aircraft losses, respectively.

Using the information available, if any, from reconnaissance and maximizing on b, we obtain

$$\max_b M(r, b) = M(r) = \int \Big\{(1 - p^2)^r \max_b \int C(b, t) \, d\phi(t) + [1 - (1 - p^2)^r] \max_b C(b, t) - (1 - p^2)Rr\Big\} \, d\phi(t),$$

where $C(b, t)$ is defined by

$$C(b, t) = t(1 - \alpha^{pb}) - (1 - p^2)Bb.$$

Finally the solution is obtained by maximizing $M(r)$ on r.

It is convenient to give the solution in terms of the following constants:

$$P = -\ln (1 - p^2),$$

$$D = -\frac{(1 - p^2)B}{p \ln \alpha},$$

$$\phi_1 = \int t \, d\phi(t),$$

$$A = D \int \ln \frac{\phi_1}{t} \, d\phi(t).$$

Further, we impose two conditions on these constants:

(i) $$\phi(D) = 0,$$

(ii) $$AP - (1 - p^2)R \geq 0.$$

The first condition states that there is no chance that the reconnaissance report will indicate that no mission should be run at all. The second condition states that there is sufficient doubt concerning the state of the target to make reconnaissance worthwhile. If the second condition fails, then no reconnaissance will be sent out.

Maximizing $M(r, b)$ on b and r, we obtain as the optimal sizes, r^*, b^*, of reconnaissance aircraft and bombers, respectively, to be dispatched:

$$r^* = 1 + \frac{1}{p} \ln \frac{AP}{R}.$$

$$b^* = \begin{cases} \dfrac{\ln\left(\dfrac{T}{D}\right)}{-p \ln \alpha} & \text{if reconnaissance reports } T \\[2ex] \dfrac{\ln\left(\dfrac{\phi_1}{D}\right)}{-p \ln \alpha} & \text{if reconnaissance does not report.} \end{cases}$$

For these optimal values of r and b we have the payoff

$$M(r^*, b^*) = \phi_1 - D - D \ln \frac{\phi_1}{D} + \frac{1}{p}[AP - (1 - p^2)R] - (1 - p^2)Rr^*.$$

If no reconnaissance were sent, then the best result would have been

$$M(b^*) = \phi_1 - D - D \ln \frac{\phi_1}{D}.$$

Example. Suppose the military worth of a bomber is the same as a reconnaissance aircraft, each equal to one unit. Prior to reconnaissance, all that is known about the target is that it is just as likely to have a value of 50 units as 5 units. Suppose each bomber can destroy 10 per cent of the worth of a target. Assume that the one-way survival probability of each aircraft is 0.8.

We have then the following given constants:

$$\begin{aligned} B &= R = 1, \\ \phi(t) &= \tfrac{1}{2}I_5(t) + \tfrac{1}{2}I_{50}(t), \\ \alpha &= 0.9, \\ p &= 0.8. \end{aligned}$$

Substituting the above constants, we obtain the following additional constants:

$$\begin{aligned} P &= -\ln(1 - 0.64) = 1.022, \\ D &= -\frac{(1 - 0.64)1}{0.8 \ln 0.9} = 4.275, \\ \phi_1 &= \int_0^{50} t\left[\frac{1}{2}\, dI_5(t) + \frac{1}{2}\, dI_{50}(t)\right] = 27.5, \\ A &= 4.275 \int \ln \frac{27.5}{t}\left[\frac{1}{2}\, dI_5(t) + \frac{1}{2}\, dI_{50}(t)\right] = 2.366, \\ \phi(D) &= \phi(4.275) = 0, \\ AP &= (1 - p^2)R = 2.058 > 0. \end{aligned}$$

The optimal numbers of reconnaissance aircraft and bombers are given by:

$$r^* = 1 + \frac{1}{1.022} \ln \frac{(2.366)(1.022)}{1} = 1.86$$

$$b^* = \begin{cases} \dfrac{\ln\left(\dfrac{5}{4.275}\right)}{-0.8 \ln 0.9} = 1.86 & \text{if reconnaissance reports 5} \\[2ex] \dfrac{\ln\left(\dfrac{50}{4.275}\right)}{-0.8 \ln 0.9} = 29.20 & \text{if reconnaissance reports 50} \\[2ex] \dfrac{\ln\left(\dfrac{27.5}{4.275}\right)}{-0.8 \ln 0.9} = 22.10 & \text{if reconnaissance does not report.} \end{cases}$$

Using these optimal strategies, the attacker receives a payoff

$$M(r^*, b^*) = 16.61.$$

If no reconnaissance had been attempted, the payoff would have been

$$M(b^*) = 15.27.$$

We may interpret the solution as follows: Two reconnaissance aircraft are to be sent out. The expected loss is $2(1 - p^2) = 0.72$ aircraft. If the reconnaissance reports that the target is worth 50 units, then 29 bombers are dispatched, causing $50(1 - 0.9^{(29)(0.8)}) = 45.67$ units of damage, with an expected loss of $29(1 - 0.64) = 10.44$ bombers. If the reconnaissance reports the target to be worth five units, then two bombers are sent out which destroy 0.77 units of the target and 0.72 bombers are expected to be lost. If the reconnaissance fails to report, then 21 bombers are dispatched, and the damage they inflict is either 42.15 or 4.22 units, or an expected 23.18 units, with an expected loss of 7.92 bombers. The probability that the reconnaissance will fail to report is $(1 - p^2)^2 = 0.13$. The probability that the reconnaissance will report 50 is the same as the probability that the reconnaissance will report 5 and is given by $\dfrac{1 - (1 - p^2)^2}{2} = 0.435$.

BIBLIOGRAPHY

American Management Association, *Top Management Decision Gaming,* May 2, 1957, American Management Association, Inc., New York, 1957.

Baldwin, Roger R., Wilbert Cantey, Herbert Maisel, and James McDermott, "The optimum strategy in blackjack," *J. Amer. Stat. Assoc.*, Vol. 51 (1956), pp. 429–439.

Bellman, Richard, and David Blackwell, "Some two-person games involving bluffing," *Proc. Nat. Acad. Sci. U.S.A.*, Vol. 35 (1949), pp. 600–605.

Beresford, R. S., and M. H. Peston, "A mixed strategy in action," *Operational Res. Q.*, Vol. 6 (1955), pp. 173–175.

Berge, Claude, "The theory of games, a new branch of mathematics," United States Line, *Paris Review Vanves,* Kapp et l'Imprimerie Illustrations Bobigny, n.d.

Berkovitz, L. D., and M. Dresher, "A game-theory analysis of tactical air war," *Operations Res.*, Vol. 7 (1959), pp. 599–620.

Bernard, Jessie, "The theory of games of strategy as a modern sociology of conflict," *Amer. J. Soc.*, Vol. LIX (1954), pp. 411–424.

Birch, B. J., "On games with almost complete information," *Proc. Cambridge Philos. Soc.*, Vol. 51 (1955), pp. 275–287.

Blackett, D. W., "Some Blotto games," *Naval Res. Log. Q.*, Vol. 1 (1954), pp. 55–60.

Blackwell, David, "Game theory," *Operations Research for Management,* J. F. McCloskey and F. N. Trefethen (eds.), Johns Hopkins Press, Baltimore, 1954, pp. 238–253.

Blackwell, David, and M. A. Girshick, *Theory of Games and Statistical Decisions,* John Wiley and Sons, Inc., New York, 1954, xi, 355 pp.

Borel, Émile, *Le Jeu, la Chance et les Théories Scientifiques Modernes,* Gallimard, 1941.

———, "On games that involve chance and the skill of the players," *Econometrica,* Vol. 21 (1953), pp. 101–115. Trans. from *Eléments de la Théorie des Probabilités,* 3e ed., Paris: Librairie Scientifique, J. Hermann, 1924, pp. 204–224.

Braithwaite, Richard B., *Theory of Games as a Tool for the Moral Philosopher,* Cambridge University Press, Cambridge, 1955, 76 pp.

Caywood, T. E., and C. J. Thomas, "Applications of game theory in fighter versus bomber combat," *J. Oper. Res. Soc.*, Vol. 3 (1955), pp. 402–411.

Chacko, George, "Certain game situations in regional economic development," *Indian J. Econ.*, Vol. 37 (1956), pp. 167–175.

Churchman, C. West, Russell L. Ackoff, and Leonard E. Arnoff, *Introduction to Operations Research,* John Wiley and Sons, Inc., New York, 1957, x, 645 pp.

Copeland, Arthur H., "John von Neumann and Oskar Morgenstern's theory of games and economic behavior," *Bull. Amer. Math. Soc.*, Vol. 51 (1945), pp. 498–504.

Dantzig, George B., "Constructive proof of the min-max theorem," *Pacific J. Math.*, Vol. 6 (1956), pp. 25–33.

Davies, Robert, "The matrix for an industrial game," *Indust. Math.*, Vol. 6 (1955), pp. 1–5.

Deutsch, Karl W., "Game theory and politics: some problems of application," *Canad. J. Econ. and Pol. Sci.*, Vol. 20 (1954), pp. 76–83.

Dresher, Melvin, "Games of strategy," *Math. Mag.*, Vol. 25 (1957), pp. 93–99.

———, "Methods of solution in game theory," *Econometrica*, Vol. 18 (1950), pp. 179–181.

Dresher, Melvin, Albert W. Tucker, and Philip Wolfe (eds.), "Contributions to the Theory of Games," *Ann. Math. Studies*, Vol. III, No. 39, Princeton University Press, Princeton, N. J., 1957, vi, 435 pp.

Fulkerson, D. R., and S. M. Johnson, "A tactical air game," *J. Oper. Res.*, Vol. 5 (1957), pp. 704–712.

Gale, David, *The Theory of Linear Economic Models*, McGraw-Hill, New York, 1960, xxi, 330 pp.

Glicksberg, Irving L., "A derivative test for finite solutions of games," *Proc. Amer. Math. Soc.*, Vol. 4 (1953), pp. 895–897.

Haywood, Col. O. G., Jr., "Military decision and the mathematical theory of games," *Air University Q.*, Rev. 4 (1950), 67 pp.

Karlin, Samuel, "The theory of infinite games," *Ann. of Math.*, Vol. 58 (1953), pp. 371–401.

———, *Mathematical Methods and Theory in Games, Programming, and Economics*, Addison-Wesley, Reading, Mass., 1959, xx, 433 pp.

Kuhn, Harold W., and Albert W. Tucker (eds.), "Contributions to the Theory of Games," *Ann. Math Studies*, Vol. I, No. 24, Princeton University Press, Princeton, N. J., 1950.

———, "Contributions to the Theory of Games," *Ann. Math. Studies*, Vol. II, No. 28, Princeton University Press, Princeton, N. J., 1953.

Loomis, Lynn H., "On a theorem of von Neumann," *Proc. Nat. Acad. Sci. U.S.A.*, Vol. 32 (1946), pp. 213–215.

Luce, R. Duncan, and Howard Raiffa, *Games and Decisions*, Introduction and Critical Survey, John Wiley and Sons, Inc., New York, 1957, xix, 509 pp.

McDonald, John, and John W. Tukey, "Colonel Blotto: a problem of military strategy," *Fortune* (June 1949), p. 102.

McKinsey, J. C. C., *Introduction to the Theory of Games*, McGraw-Hill, New York, 1952, x, 371 pp.

Morgenstern, Oskar, "The theory of games," *Sci. Amer.*, Vol. 180 (1949), pp. 22–25.

von Neumann, John, "A numerical method to determine optimum strategy," *Naval Res. Log. Q.*, Vol. 1 (1954), pp. 109–115.

———, "Zur Theorie der Gesellschaftsspiele," *Math. Ann.*, Vol. 100 (1928), pp. 295–320.

von Neumann, John, and Oskar Morgenstern, *Theory of Games and Economic Behavior*, Princeton University Press, Princeton, 1944, xviii, 625 pp.

Nikaidô, Hukukane, "On von Neumann's minimax theorem," *Pacific J. Math.*, Vol. 4 (1954), pp. 65–72.

Shapley, Lloyd S., "Stochastic games," *Proc. Nat. Acad. Sci. U.S.A.*, Vol. 39 (1953), pp. 1095–1100.

Shapley, Lloyd S., and Martin Shubik, "Method for evaluating the distribution of power in a committee system," *Amer. Pol. Sci. Rev.*, Vol. 48 (1954), pp. 787–792.

Shubik, Martin, "A game theorist looks at the antitrust laws and the automobile industry," *Stanford Law Review*, Vol. 8 (1956), pp. 594–630.

———, "Information, theories of competition and the theory of games," *J. Pol. Econ.*, Vol. 60 (1953), pp. 142–150.

Vajda, S., *The Theory of Games and Linear Programming*, Methuen and Co., London, John Wiley and Sons, Inc., New York, 1956, 106 pp.

Williams, John, *The Compleat Strategyst*, McGraw-Hill, New York, 1954, xiii, 234 pp.

INDEX

F

G

H

I

L

M

A CATALOG OF SELECTED

DOVER BOOKS

IN SCIENCE AND MATHEMATICS

ROTARY-WING AERODYNAMICS, W.Z. Stepniewski. Clear, concise text covers aerodynamic phenomena of the rotor and offers guidelines for helicopter performance evaluation. Originally prepared for NASA. 537 figures. 640pp. 6⅛ × 9¼.
64647-5 Pa. $14.95

DIFFERENTIAL GEOMETRY, Heinrich W. Guggenheimer. Local differential geometry as an application of advanced calculus and linear algebra. Curvature, transformation groups, surfaces, more. Exercises. 62 figures. 378pp. 5⅜ × 8½.
63433-7 Pa. $7.95

INTRODUCTION TO SPACE DYNAMICS, William Tyrrell Thomson. Comprehensive, classic introduction to space-flight engineering for advanced undergraduate and graduate students. Includes vector algebra, kinematics, transformation of coordinates. Bibliography. Index. 352pp. 5⅜ × 8½. 65113-4 Pa. $8.00

A SURVEY OF MINIMAL SURFACES, Robert Osserman. Up-to-date, in-depth discussion of the field for advanced students. Corrected and enlarged edition covers new developments. Includes numerous problems. 192pp. 5⅜ × 8½.
64998-9 Pa. $8.95

ANALYTICAL MECHANICS OF GEARS, Earle Buckingham. Indispensable reference for modern gear manufacture covers conjugate gear-tooth action, gear-tooth profiles of various gears, many other topics. 263 figures. 102 tables. 546pp. 5⅜ × 8½.
65712-4 Pa. $11.95

SET THEORY AND LOGIC, Robert R. Stoll. Lucid introduction to unified theory of mathematical concepts. Set theory and logic seen as tools for conceptual understanding of real number system. 496pp. 5⅜ × 8¼. 63829-4 Pa. $8.95

A HISTORY OF MECHANICS, René Dugas. Monumental study of mechanical principles from antiquity to quantum mechanics. Contributions of ancient Greeks, Galileo, Leonardo, Kepler, Lagrange, many others. 671pp. 5⅜ × 8½.
65632-2 Pa. $14.95

FAMOUS PROBLEMS OF GEOMETRY AND HOW TO SOLVE THEM, Benjamin Bold. Squaring the circle, trisecting the angle, duplicating the cube: learn their history, why they are impossible to solve, then solve them yourself. 128pp. 5⅜ × 8½.
24297-8 Pa. $3.95

MECHANICAL VIBRATIONS, J.P. Den Hartog. Classic textbook offers lucid explanations and illustrative models, applying theories of vibrations to a variety of practical industrial engineering problems. Numerous figures. 233 problems, solutions. Appendix. Index. Preface. 436pp. 5⅜ × 8½. 64785-4 Pa. $8.95

CURVATURE AND HOMOLOGY, Samuel I. Goldberg. Thorough treatment of specialized branch of differential geometry. Covers Riemannian manifolds, topology of differentiable manifolds, compact Lie groups, other topics. Exercises. 315pp. 5⅜ × 8½.
64314-X Pa. $6.95

HISTORY OF STRENGTH OF MATERIALS, Stephen P. Timoshenko. Excellent historical survey of the strength of materials with many references to the theories of elasticity and structure. 245 figures. 452pp. 5⅜ × 8½. 61187-6 Pa. $10.95

GEOMETRY OF COMPLEX NUMBERS, Hans Schwerdtfeger. Illuminating, widely praised book on analytic geometry of circles, the Moebius transformation, and two-dimensional non-Euclidean geometries. 200pp. 5⅝ × 8¼.
63830-8 Pa. $6.95

MECHANICS, J.P. Den Hartog. A classic introductory text or refresher. Hundreds of applications and design problems illuminate fundamentals of trusses, loaded beams and cables, etc. 334 answered problems. 462pp. 5⅜ × 8½. 60754-2 Pa. $8.95

TOPOLOGY, John G. Hocking and Gail S. Young. Superb one-year course in classical topology. Topological spaces and functions, point-set topology, much more. Examples and problems. Bibliography. Index. 384pp. 5⅜ × 8¼.
65676-4 Pa. $7.95

STRENGTH OF MATERIALS, J.P. Den Hartog. Full, clear treatment of basic material (tension, torsion, bending, etc.) plus advanced material on engineering methods, applications. 350 answered problems. 323pp. 5⅜ × 8½. 60755-0 Pa. $7.50

ELEMENTARY CONCEPTS OF TOPOLOGY, Paul Alexandroff. Elegant, intuitive approach to topology from set-theoretic topology to Betti groups; how concepts of topology are useful in math and physics. 25 figures. 57pp. 5⅜ × 8½.
60747-X Pa. $2.95

ADVANCED STRENGTH OF MATERIALS, J.P. Den Hartog. Superbly written advanced text covers torsion, rotating disks, membrane stresses in shells, much more. Many problems and answers. 388pp. 5⅜ × 8½. 65407-9 Pa. $9.95

COMPUTABILITY AND UNSOLVABILITY, Martin Davis. Classic graduate-level introduction to theory of computability, usually referred to as theory of recurrent functions. New preface and appendix. 288pp. 5⅜ × 8½. 61471-9 Pa. $6.95

GENERAL CHEMISTRY, Linus Pauling. Revised 3rd edition of classic first-year text by Nobel laureate. Atomic and molecular structure, quantum mechanics, statistical mechanics, thermodynamics correlated with descriptive chemistry. Problems. 992pp. 5⅜ × 8½. 65622-5 Pa. $18.95

AN INTRODUCTION TO MATRICES, SETS AND GROUPS FOR SCIENCE STUDENTS, G. Stephenson. Concise, readable text introduces sets, groups, and most importantly, matrices to undergraduate students of physics, chemistry, and engineering. Problems. 164pp. 5⅜ × 8½. 65077-4 Pa. $5.95

THE HISTORICAL BACKGROUND OF CHEMISTRY, Henry M. Leicester. Evolution of ideas, not individual biography. Concentrates on formulation of a coherent set of chemical laws. 260pp. 5⅜ × 8½. 61053-5 Pa. $6.00

THE PHILOSOPHY OF MATHEMATICS: An Introductory Essay, Stephan Körner. Surveys the views of Plato, Aristotle, Leibniz & Kant concerning propositions and theories of applied and pure mathematics. Introduction. Two appendices. Index. 198pp. 5⅜ × 8½. 25048-2 Pa. $5.95

THE DEVELOPMENT OF MODERN CHEMISTRY, Aaron J. Ihde. Authoritative history of chemistry from ancient Greek theory to 20th-century innovation. Covers major chemists and their discoveries. 209 illustrations. 14 tables. Bibliographies. Indices. Appendices. 851pp. 5⅜ × 8½. 64235-6 Pa. $17.95

DE RE METALLICA, Georgius Agricola. The famous Hoover translation of greatest treatise on technological chemistry, engineering, geology, mining of early modern times (1556). All 289 original woodcuts. 638pp. 6¾ × 11.
60006-8 Clothbd. $17.95

SOME THEORY OF SAMPLING, William Edwards Deming. Analysis of the problems, theory and design of sampling techniques for social scientists, industrial managers and others who find statistics increasingly important in their work. 61 tables. 90 figures. xvii + 602pp. 5⅜ × 8½. 64684-X Pa. $15.95

THE VARIOUS AND INGENIOUS MACHINES OF AGOSTINO RAMELLI: A Classic Sixteenth-Century Illustrated Treatise on Technology, Agostino Ramelli. One of the most widely known and copied works on machinery in the 16th century. 194 detailed plates of water pumps, grain mills, cranes, more. 608pp. 9 × 12.
25497-6 Clothbd. $34.95

LINEAR PROGRAMMING AND ECONOMIC ANALYSIS, Robert Dorfman, Paul A. Samuelson and Robert M. Solow. First comprehensive treatment of linear programming in standard economic analysis. Game theory, modern welfare economics, Leontief input-output, more. 525pp. 5⅜ × 8½. 65491-5 Pa. $13.95

ELEMENTARY DECISION THEORY, Herman Chernoff and Lincoln E. Moses. Clear introduction to statistics and statistical theory covers data processing, probability and random variables, testing hypotheses, much more. Exercises. 364pp. 5⅜ × 8½. 65218-1 Pa. $8.95

THE COMPLEAT STRATEGYST: Being a Primer on the Theory of Games of Strategy, J.D. Williams. Highly entertaining classic describes, with many illustrated examples, how to select best strategies in conflict situations. Prefaces. Appendices. 268pp. 5⅜ × 8½. 25101-2 Pa. $5.95

MATHEMATICAL METHODS OF OPERATIONS RESEARCH, Thomas L. Saaty. Classic graduate-level text covers historical background, classical methods of forming models, optimization, game theory, probability, queueing theory, much more. Exercises. Bibliography. 448pp. 5⅜ × 8¼. 65703-5 Pa. $12.95

CONSTRUCTIONS AND COMBINATORIAL PROBLEMS IN DESIGN OF EXPERIMENTS, Damaraju Raghavarao. In-depth reference work examines orthogonal Latin squares, incomplete block designs, tactical configuration, partial geometry, much more. Abundant explanations, examples. 416pp. 5⅜ × 8¼.
65685-3 Pa. $10.95

THE ABSOLUTE DIFFERENTIAL CALCULUS (CALCULUS OF TENSORS), Tullio Levi-Civita. Great 20th-century mathematician's classic work on material necessary for mathematical grasp of theory of relativity. 452pp. 5⅜ × 8½.
63401-9 Pa. $9.95

VECTOR AND TENSOR ANALYSIS WITH APPLICATIONS, A.I. Borisenko and I.E. Tarapov. Concise introduction. Worked-out problems, solutions, exercises. 257pp. 5⅜ × 8¼. 63833-2 Pa. $6.95